T0138433

A MARTIAN
STRANDED
ON EARTH

A MARTIAN STRANDED ON EARTH

ALEXANDER BOGDANOV,

BLOOD TRANSFUSIONS,

AND PROLETARIAN

SCIENCE

Nikolai Krementsov

THE UNIVERSITY OF CHICAGO PRESS

CHICAGO AND LONDON

NIKOLAI KREMENTSOV is associate professor at the Institute for the History and Philosophy of Science and Technology at the University of Toronto. He has contributed to several scholarly journals and edited volumes, and his previous books include *Stalinist Science*, *The Cure*, and *International Science*.

The University of Chicago Press, Chicago 60637
The University of Chicago Press, Ltd., London
© 2011 by The University of Chicago
All rights reserved. Published 2011.
Printed in the United States of America

20 19 18 17 16 15 14 13 12 11   1 2 3 4 5

ISBN-13: 978-0-226-45412-2 (cloth)
ISBN-10: 0-226-45412-6 (cloth)

Library of Congress Cataloging-in-Publication Data
Krementsov, N. L.
    A Martian stranded on Earth : Alexander Bogdanov,
    blood transfusions, and proletarian science / Nikolai
    Krementsov.
        p. cm.
        Includes bibliographical references and index.
        ISBN-13: 978-0-226-45412-2 (cloth : alk. paper)
        ISBN-10: 0-226-45412-6 (cloth : alk. paper)
    1. Bogdanov, A. (Aleksandr), 1873–1928.
    2. Biology—Soviet Union—History.
    3. Blood—Transfusion.
    4. Science—Soviet Union—History. I. Title.
    QH305.2S63K74 2011   570.947′0904—dc22   2010039373

♾ This paper meets the requirements of ANSI/NISO Z39.48-1992
(Permanence of Paper).

To my friends

We are born to make a fairy tale into reality.

Pavel German and Iulii Khait,
"Ever higher" (circa 1924)

Whatever the judgment of our elementally dramatic epoch by the future historian will be—and, undoubtedly, his assessment will differ profoundly from a contemporary one—in any case, he will note and recognize the *unprecedented audacity in the formulation of problems* as one of the best manifestations of this epoch's cultural process.

Alexander Bogdanov,
Struggle for Viability (1927)

Contents

Illustrations

Note on Translation and Transliteration

Although some of the original Russian sources I cite are available in English, all the translations in the book are my own. Throughout the book, I use the Library of Congress's transliteration system, except for the commonly adopted spellings of well-known names—for example, "Trotsky" and "Metchnikoff" instead of "Trotskii" and "Mechnikov."

Genealogy and Acknowledgments

Until about a year ago, I had never thought that I would write a book about Alexander Alexandrovich Malinovskii-Bogdanov, even though I had known of him for nearly forty years. I first encountered his name in 1973, when, as a freshman at Rostov-on-Don University's Biology School, I took the obligatory course "The History of the Communist Party of the Soviet Union." In the fat gray textbook of nearly eight hundred pages, Bogdanov's name was mentioned twice. The first time was in the list of participants in the Third Party Congress held in 1905 in London. The second time Bogdanov appeared in a chapter entitled "V. I. Lenin's Defense and Development of Marxist Philosophy" as one of several "bourgeois *intelligenty*," who "had attacked the fundamental principles of Marxist philosophy" in the aftermath of the first Russian revolution of 1905–6 (Ponomarev 1973, 74, 122). The next year, taking another obligatory course on dialectical materialism, I had to read *Materialism and Empiriocriticism*, Lenin's "foundational" treatise on the subject. In the process, I learned that it had been written specifically to undermine Bogdanov's "reactionary philosophy." Bogdanov's name—accompanied by "appropriate" epithets—figured on practically every other page of Lenin's four-hundred-page-long work. Bogdanov's own treatise on "empiriomonism," which Lenin had been lambasting throughout the text, however, was not on the course's reading list, and I could not even glimpse its actual contents from partial quotations scattered within Lenin's "critique." Nor could I guess that Bogdanov had written a similarly acerbic response to Lenin's tract: it was never mentioned in any of the course readings. Needless to say, I forgot Bogdanov's name as soon as I passed (with some difficulty) the exams: at the time, I was much more interested in neurocybernetics than in memorizing some obscure names from the history of the Communist Party or dialectical materialism.

Yet, some years later, I encountered Bogdanov's name again. In 1990, during the heyday of Mikhail Gorbachev's perestroika, Politizdat—a publishing house run by the Communist Party's Central Committee, which had issued the textbook on the history of the party that I had read as a student, as well as Lenin's *Materialism and Empiriocriticism*—released a volume of Bogdanov's selected works entitled *Issues of Socialism*. I probably would never have noticed the book—"issues of socialism" not being high on my interest list—had it not been for a friend who told me that, along with a series of articles, the volume included two science-fiction (SF) novels about Mars and its inhabitants, one entitled *Red Star*, another *Engineer Menni*. A longtime SF fan, I bought the book and, skipping the articles altogether, read the novels. Alas, conditioned by the classics of Western and Soviet SF, ranging from H. G. Wells, Isaac Asimov, and Ray Bradbury to Alexander Beliaev, Ivan Efremov, and Boris and Arkadii Strugatskiis, I found Bogdanov's Martian novels dull and unimaginative by comparison. In fact, bored to tears, I could not even finish the second one and left the volume to gather dust on my bookshelf.

By that time, I had lost interest in neurocybernetics and had moved on to studying the history of science—to be exact, the history of Russian biological and biomedical sciences. Indeed, earlier that very year, I had defended my doctoral dissertation on the history of interactions between behavioral and evolutionary research and had begun working on my first monograph about the Soviet science system during the Stalin era. Since then—for the last twenty years—I have been writing on various aspects of the history of Russian biomedical sciences, and through these years, Bogdanov's name has periodically lurked somewhere on the periphery of my investigations.

Five years ago, I started a large project on the effects of the 1917 Bolshevik Revolution in experimental biology and experimental medicine. I wanted to look at intellectual and institutional developments in several disciplines and research programs, from endocrinology and eugenics to biophysics and gerontology, that had flourished in Russia in the aftermath of the revolution. One such discipline was to be "blood science," for I had a deep personal interest in this particular science. According to my family's lore, in the early 1930s, the prominent Tashkent surgeon Valentin Voino-Iasenetskii saved a little boy with a very bad case of septicemia from inevitable death by . . . a blood transfusion—there were no antibiotics or even sulfa drugs in those days. That little boy grew up to become my father, and I, of course, was eager to find out more about the history of blood transfusions in the Soviet Union. To my surprise, a preliminary review of the secondary literature suddenly brought Bogdanov from the periphery to the forefront of my research. I quickly learned that he had played a major role in the history of Soviet blood studies and that the Moscow Scientific-Research Institute of

Blood Transfusion still bears the name of its founder and first director, Bogdanov! I blew off the thick layer of dust from the 1990 Bogdanov volume still on my shelf and for the first time read it from cover to cover. Although some of the articles included in the volume dealt with Marxist analyses of what science is and ought to be under socialism, none of them had anything to do with "blood science." But one of the SF novels, *Red Star,* did mention "blood exchanges" as a means of "rejuvenation" practiced by the inhabitants of the "Red Planet." Marxism, SF, and blood research? It did not make much sense. Intrigued, I launched a full-scale hunt for Bogdanov's materials in Russian archives, libraries, and secondhand bookstores as part of my ongoing research on the history of 1920s biomedical sciences in Russia.

For four years, I collected archival documents, journal and newspaper articles, books, and photographs; visited museums; and read the extensive secondary literature on the history of experimental biology and experimental medicine, along the way writing up various pieces of my research pertaining to specific disciplines and research directions. About fifteen months ago, I began drafting what I thought would be a fifty-page chapter on blood transfusions in a book about biomedical sciences and the Bolshevik Revolution. Quite unexpectedly, the draft quickly grew to twice the projected length. It was obviously far too long to be published as a chapter in the book or as a separate article, but I was reluctant to cut it in half—the story was simply too fascinating and far too complex to be told in fifty pages. I had a vague feeling that it merited a book in its own right but was unsure whether I should take time away from my half-finished "big" book to write it up. On a whim, in May 2009, I sent the draft to Karen Merikangas Darling, an editor at the University of Chicago Press. It was Karen's enthusiasm that convinced me that my "instinct" had been correct: there was a book to be written. During the summer, while my wife and daughter enjoyed fresh scallops and tuna steaks in Cape Cod, I holed up in our house in Toronto and, in a burst of inspiration, wrote the book in just five weeks.

As is always the case with historical work, this book bears the mark of numerous individuals. The help of Aleksandra Bekasova, Dmitrii Derviz, Igor Kozyrin, Galina Savina, Aleksandr Sledkov, and Marina Sorokina in collecting documents from archives, libraries, and museums in Moscow and St. Petersburg was indispensable. John B. Appleby, Yaroslav Ivanov, and Elena Maltseva provided much-appreciated assistance in Toronto, carrying out the burden of surveying and copying relevant items from the secondary literature. The staffs of numerous libraries and archives were extremely helpful, including the Archive of the Russian Academy of Sciences, the Russian State Archive of Socio-Political History, the State

Archive of the Russian Federation (Moscow), the U.S. National Library of Medicine (Bethesda, Maryland), the Libraries of the University of Toronto, the Library of the Russian Academy of Sciences (St. Petersburg), the National Library of Russia (St. Petersburg), the Museum of the History of Medicine of the Sechenov Moscow Medical Academy, and the Military Medicine Museum of the Russian Federation Ministry of Defense (St. Petersburg).

Certain ideas developed in this book were presented at a conference on the history of Soviet medicine in May 2005, in Gregynog (Wales, United Kingdom); at a research seminar of The Johns Hopkins University's Institute for the History of Medicine (Baltimore, Maryland) in February 2009; and in the "Human Biology and Human Destiny" and "The Biology of Death and Immortality" courses that I taught at the University of Toronto from 2005 to 2009, and I would like to thank the participants for inspiring questions and stimulating discussions. I am indebted to Mark B. Adams, Anne-Emanuelle Birn, Susan G. Solomon, and Daniel P. Todes, each of whom read the entire manuscript, for their thoughtful comments, which helped improve the book in many significant ways. I am also grateful to the anonymous reviewers of the manuscript for the University of Chicago Press for their helpful suggestions. I would like to express my deepest gratitude to Karen Merikangas Darling, who became a virtual "midwife" to this book: it was her interest and encouragement that led me to write it in the first place, and her unfailing support proved crucial to bringing this project to fruition.

Financial support for research leading to this book came from the Social Sciences and Humanities Research Council of Canada and from the U.S. National Institutes of Health (through National Library of Medicine publication grant G13 LM008632), for which I am extremely grateful. I would also like to thank the University of Toronto's Institute of the History and Philosophy of Science and Technology for supporting my work and for providing a generous subvention grant for its publication.

Naturally, I alone am responsible for any mistakes and misinterpretations.

St. Petersburg–Geneva, April–May 2010

1 Revolutions

THE BIG SCIENCE OF VISIONARY BIOLOGY

On April 4, 1926, *Izvestiia,* the Soviet Union's most widely circulated newspaper, carried a long article about a "scientific research institute of blood transfusion" that had been recently created in Moscow by the People's Commissariat of Health Protection (Narkomzdrav), the country's highest agency in charge of medicine and public health. Signed by the director of the new institution, a certain "A. Bogdanov," the article detailed the tasks and prospects of the "world's first" research institute devoted exclusively to studies of blood transfusion. Bogdanov asserted that, following in the footsteps of their "perpetual teachers"— German physicians—Russian doctors had lagged, "and criminally so," far behind their French, British, and, particularly, U.S. colleagues in adopting this "lifesaving procedure." Since "blood transfusions have enormous scientific and social import," Bogdanov continued, the Soviet Union must catch up with foreign advances, and his institute had been established to further precisely this goal. Bogdanov claimed that Narkomzdrav had assigned him the task of organizing the institute because he understood both the "social-practical and scientific importance" of blood transfusions. He described his own long-standing interest in the procedure, which had led him to formulate a concept of "physiological collectivism"—the increase of the "viability" of individual organisms through regular blood exchanges among them. Bogdanov reported that in early 1924 he had organized a research group to study "physiological collectivism," which became "the first in the country to practice the modern technique of blood transfusion" and which to date had performed "about ten" blood exchanges. The results—because of the small number of cases—"do not allow [me] to draw decisive conclusions," Bogdanov admitted, "but they are certainly encouraging and call for the continuation of research" under the auspices of the new institute (Bogdanov 1926, 5).

The *Izvestiia* article probably appeared deeply puzzling to its readers. Those who were professionally connected with experimental biology and experimental medicine were likely delighted to learn that Narkomzdrav had finally established a special institute to study one of the most exciting subjects in their field. But they had never before heard of "Dr. Bogdanov," his research group, or his concept of "physiological collectivism." None of the country's numerous medical and biological journals had ever carried any publication by "A. Bogdanov" on his ideas and experiments, even though they had published a variety of research articles, literature surveys, and professional debates devoted to blood groups and blood transfusions. Indeed, in the preceding years, several Russian surgeons, including Vladimir Shamov, Nikolai Elanskii, Iakov Bruskin, and Erik Gesse, had developed original transfusion techniques and apparatuses, produced their own standard sera for blood typing, performed transfusions, reported their results to various professional meetings, and published accounts of their work in leading professional and popular periodicals. The readers would perhaps have expected that Narkomzdrav would have appointed one of these well-respected doctors—and not someone completely unknown to professionals—to head a research institute of blood transfusion. After all, this was a customary practice: the country's most eminent physicians and scientists stood at the helm of the nearly forty research institutions Narkomzdrav had already established by the spring of 1926.

Other readers—mostly among students and recent graduates of various Soviet institutions of higher learning—were also likely confused. They had known one "A. Bogdanov," the author of the country's most popular textbook on Marxist political economy and of numerous articles and books on Marxist philosophy—two obligatory subjects in their curricula. Some of them might have also known "A. Bogdanov" as a leader of the Proletkult (proletarian culture) movement and the founder of "tectology," as he named his concept of "universal organizational science." But even if they had noticed certain lexical parallels and similarities in ideas between the article's "physiological collectivism" and "A. Bogdanov's" philosophical writings, the distance between physiology and philosophy was simply too great, and they likely would have thought that the article was written by a different Bogdanov. After all, it is a fairly common Russian surname.

A certain part of the *Izvestiia* readership—predominantly among the younger generation—perhaps recognized both the author and the ideas presented in the article. They had read about "blood exchanges" before—in a popular science-fiction (SF) novel entitled *Red Star*, authored by "A. Bogdanov." The novel had first appeared in 1908, and since the Bolshevik Revolution of 1917, it had been reprinted no fewer than six times, performed as a theatrical play, and translated into several languages of the Soviet Union. But in the novel, it was on Mars that blood exchanges were used to extend the life span of the red planet's inhabitants. Had Martians invaded Narkomzdrav? For these readers, the fact that Narkomzdrav had organized a special institute to study blood exchanges in Moscow and appointed the man who had first envisioned their application in an SF novel resonated well with their firmly held belief—aptly expressed in a popular song of the time—that "we are born to make a fairy tale into reality." They likely did not know that "A. Bogdanov" had actually held a medical diploma and hence had at least the formal qualifications to direct a Narkomzdrav research establishment. For them, his appointment perhaps signaled that a basic principle of "proletarian" science enunciated in Bogdanov's novels—that anyone can master the "secrets" of science—had triumphed in Soviet Russia as it had on Bogdanov's fictional Mars.

Very few among the readers of the *Izvestiia* article likely suspected and still fewer actually knew that these three "different" Bogdanovs—the scientist-physician, the Marxist, and the litterateur—were in fact one and the same man. "A. Bogdanov" was a pen name of Alexander Alexandrovich Malinovskii. Born in 1873, Malinovskii initially studied natural sciences and later medicine. In the 1900s, he became an influential Marxist propagandist and theoretician, one of the leaders of the Bolshevik Party, and, at the same time, a writer of popular science-fiction novels. The confusing "trinity" of Bogdanov's persona has been perpetuated in the historical record, and it is a purpose of this work to find a unity, or, using Bogdanov's own terminology, a "universal organizational principle" in the life and works of this versatile and multitalented man.

ALEXANDER BOGDANOV

Vladimir Lenin's rival in the leadership of the Bolshevik Party during its formative years in the early 1900s, Malinovskii-Bogdanov has loomed large in the works of historians of Russia over the last few decades.[1] They have reprinted and translated many of his writings, produced several biographies, held special conferences, and published innumerable articles analyzing Bogdanov's role in Russian politics, philosophy, economics, literature, science, and medicine. Commentators have variously hailed Bogdanov as the classic of Soviet science fiction, as a forebear of systems theory and cybernetics, as an original and influential Marxist economist and philosopher, as the chief ideologue of the Proletkult movement, and as the founder of the Soviet system of blood research and services.

Yet, the three "personas" of Bogdanov—the Marxist, the novelist, and the scientist-physician—never really cohere in these accounts.[2] To be sure, Bogdanov's "contribution to Russian medicine" has figured prominently in practically all retellings of his life story, but largely due to its tragic end. Two years after he had come to head the Institute of Blood Transfusion, Bogdanov died after exchanging nearly one liter of blood with a student suffering from tuberculosis, in an attempt to prove experimentally his concept of "physiological collectivism." Combined with the lack of detailed knowledge of the early history of blood transfusions in Soviet Russia, this tragedy has certainly colored hagiographic accounts of Bogdanov's life and works, and it has provided the basis for recurrent speculations on whether Bogdanov's death was actually a suicide or even a homicide. Yet, in the numerous analyses of his life, Bogdanov's labors as a practicing scientist and director of a research institution have been all but ignored.[3]

At the same time, available histories of Soviet blood research have long praised Bogdanov's "lifelong interest in blood transfusions," his "pioneering ideas," and his "considerable organizational efforts," which allegedly played a decisive role in the introduction of blood transfusions into Soviet clinical practice and which were epitomized in the establishment of "the world's first" research institute for that very purpose.[4] Yet these histories have completely neglected Bogdanov's philosophical and fictional writings about science and, most important, how these writings might have informed and shaped his own "blood science."

Similarly, students of Bogdanov's literary works have emphasized the obvious influence his Marxist convictions exerted on his fiction, analyzing in detail what they have termed the "first Bolshevik utopia."[5] They have drawn parallels and shown distinctions between Bogdanov's novels and the classical utopias of Plato, Thomas More, Francis Bacon, and Tommaso Campanella. They have traced in his novels themes, styles, motifs, characters, and composition techniques common to the classic works of the nascent genre of science fiction produced by H. G. Wells, Edgar Rice Burroughs, and many other SF pioneers in Britain, France, Germany, Russia, and the United States.[6] But they have never seriously explored the interrelations between Bogdanov's particular scientific/medical expertise, experiences, and expectations, on the one hand, and his fiction, on the other.[7] To give but one example, in the most thorough and informed examination to date of the interplay between utopian visions and reality in postrevolutionary Russia, even though Bogdanov's name is one of the most frequently mentioned (second only to Lenin's), his science is not mentioned at all (Stites 1989).

In what follows, I show that the three different vocations Bogdanov actively pursued throughout his life—Marxism, literature, and science—intertwined, interacted, and mutually reinforced one another. I suggest that a key "organizational principle" that informed and shaped Bogdanov's varied activities was his concept of *proletarian science:* what science is and what it is supposed to be in a socialist society. In its various incarnations and at various stages of its development, this concept provided a common frame to the majority of Bogdanov's philosophical works. It lay at the center of his SF novels. It infused his writings on and practical work in the Proletkult. It shaped his personal scientific interests, affecting profoundly the directions, practices, and interpretations of his actual research. It provided guidance and inspiration for his experiments with blood exchanges. And, as we will see, it might well have predetermined the ultimate failure of his research program.

BLOOD TRANSFUSIONS

At the same time, careful examination of Bogdanov's work in the context of contemporary experimental biology and medicine—and of research on blood transfusion more specifically—sheds much-needed light on the early history of blood research and blood services in Soviet Russia. In the last few decades, the history of blood transfusion has attracted considerable attention in both academic and lay circles.[8] Scholars have examined the developments of ideas, techniques, and institutions pertinent to the introduction and spread of this lifesaving procedure in Britain, Canada, France, Germany, and the United States, spurred during the last century by the discovery of blood groups.[9] They have detailed the transfer of ideas and techniques among practitioners in various countries and have emphasized the role that the two world wars played in the diffusion and popularization of blood transfusions.[10] Alas, most of these studies have neglected the history of blood transfusions in Russia, creating a marked imbalance in their comparative analyses. Those few Western scholars who have touched upon certain events of that history—and especially, the role of Bogdanov in these events—have often repeated uncritically a few hagiographic accounts that circulated in Soviet medical histories.[11] Even though Russian scholars have produced a sizable literature documenting the history of "blood science" in their country,[12] this literature has never been fully integrated with and into the Western scholarship, which has led to the perpetuation of certain myths and misconceptions in both Russian and Western histories.[13]

On the basis of newly available archival and published materials, this book aims to reassess Bogdanov's role in the creation of the Soviet system of blood research and services and to place his ideas, experiments, and the institute he created in the intellectual, institutional, and medical contexts of the interwar period. Why did Narkomzdrav decide to establish the Institute of Blood Transfusion in the spring of 1926? Why was Bogdanov—a man better known as a philosopher, economist, and writer, but not as a physician, and least of all a scientist—appointed the director of this new institute? What role did Bogdanov's vision of "physiological collectivism" and his institute play in the development of "blood science" in the Soviet Union? The search for answers to these questions forms the core narrative of this book.

Nonetheless, this book is neither a new biography of Alexander Bogdanov nor simply a new analysis of his philosophical, political, and sociological views, even though both deserve thorough, book-length treatments. Much of this volume deals with blood research in Russia, but it is not a history of Soviet blood services and blood transfusions either. Rather, in combining elements of the three subjects, this book examines the development of biomedical sciences in 1920s Russia through the lens of Bogdanov's involvement with blood studies and the institute of blood transfusion, on the one hand, and of his fictional and philosophical writings on science, on the other. In my opinion, Bogdanov's case provides a unique opportunity to understand the development of science in Russia during the first decades of the Bolshevik regime.

BIG SCIENCE, VISIONARY BIOLOGY, AND SF

A historical coincidence influenced dramatically the development of Soviet science: just as Russia was going through its brutal political revolutions, science was undergoing its own "little" revolution—a revolution of scale. At the start of the twentieth century, science began its transformation from a small-size endeavor of individual researchers (and their students), who made their own simple instruments and often financed their own pursuits, into a huge industry-like enterprise that involved large, specialized institutions, hundreds of workers, complex machinery, and ever more resources. Scientists all over the world desperately sought patrons and partners to provide the support and funding necessary for this emerging "big science." In Russia, they found such a partner—the Bolshevik state.

No patrons were more willing or more enthusiastic in their support of science than the Bolsheviks. In just two decades after the revolution, the combined efforts of scientists and the new Bolshevik government transformed Russia from a modest province of world science into one of its great centers. Each partner had its own vision of this joint venture, each had something to gain from it, and each had a price to pay. Various scholars have explored the mechanics and dynamics of the alliance between Russian scientists and the Bolshevik state, identified the talking partners and the languages they spoke, and analyzed the institutional structures and professional cultures that emerged as a result of this developing symbiosis.[14] They have examined in detail the multilayered interactions between science and the state in the 1920s through the 1930s, which paved the way for the "little" revolution—the transition from "little science" to "big science"—in Soviet Russia.[15]

The Russian revolutions also coincided with another scientific revolution, one that affected not all natural sciences but only a particular subset, one that nowadays is called the life sciences but at the time was represented by the two interconnected and overlapping fields of experimental medicine and experimental biology.[16] The very names of these fields allude to the essence of this third "mini" revolution: the introduction of experimental methods—largely borrowed from physics and chemistry—into the studies of life, death, and disease. Begun in the last two decades of the nineteenth century, this revolution reached its apex in the 1910s and 1920s.[17] Armed with the new experimental methods, numerous researchers around the world enthusiastically attacked the mysteries of basic life processes and their pathological changes, including metabolism, reproduction, nervous and endocrine regulation, cell division, psychological and behavioral patterns, variability, immunity, growth, and heredity. The advances of experimental research were quickly deployed in various branches of medicine, leading to new diagnostic techniques, therapeutic treatments, surgical procedures, and preventive measures. In the first decades of the twentieth century, successes in tissue and organ transplantation; blood transfusion; serodiagnostics; hormone, vitamin, and sera therapies; cell and tissue cultivation; and in deciphering the basic mechanisms of heredity and embryonic development generated a euphoric vision: science could control life, death, and disease.[18] Captivated by this "visionary biology," as Mark B. Adams (2000) has aptly termed it, many scientists around the world came to believe that experimental techniques could provide them with the means to improve dramatically the wellbeing of humanity, control the spread of diseases, extend individual life spans, manipulate human reproduction and evolution, and maybe even realize a centuries-old dream of immortality.[19] Russian scientists were no exception.

As elsewhere around the world, in Russia, the development of experimental biology and medicine had begun in the last decades of the nineteenth century and was epitomized in 1890 by the establishment of the Imperial Institute of Experimental Medicine in the country's capital, St. Petersburg. In subsequent years, a large number of Russian scientists became engaged in research on new, exciting biomedical subjects. Two Nobel Prizes awarded to Ivan Pavlov (1904) and Elie Metchnikoff (1908) symbolized the international recognition of Russian contributions. But despite the considerable efforts of many scientists, the institutional development of Russian experimental biology and medicine lagged behind their intellectual advances.

The situation changed dramatically after the 1917 Bolshevik Revolution, which fostered a virtual explosion of research in numerous fields of "visionary biology" and resulted in the creation of large, specialized research institutions characteristic of "big science." Within just one decade, a number of disciplines and research directions previously underdeveloped or nonexistent in Russia became quickly institutionalized. Journals, societies, conferences, research institutes, and teaching departments in such new disciplines as endocrinology, genetics, hematology, immunology, zoopsychology, experimental cytology and embryology, biochemistry, eugenics, dietetics, the physiology of "higher nervous processes," social hygiene, "psychotechnology," biophysics, pedology (the study of childhood), and experimental psychophysiology proliferated throughout the country.

Of course, in certain ways, these intellectual and institutional developments mirrored similar processes unfolding during the same time elsewhere in Europe and the Americas. During the 1920s, investigators around the globe conducted similar research and strove to build specialized institutions to further their studies of new, exciting biomedical subjects.[20] What made the Russian case special was that after 1917, science in Russia became an exclusively state-sponsored enterprise. If elsewhere in the world the drive to institutionalize and advance the new directions in biomedical research was to a large degree promoted and supported by private individuals, universities, foundations, and corporations, in Russia the sole patron was the state.[21]

How did Russian scientists win the ear of various state agencies and agents who fostered the unprecedented expansion of research in experimental biology and medicine in new Soviet Russia? And why was the Soviet leadership willing to commit resources of the new state (scarce as they were) to building large research institutions in these new fields? The ideal of controlling and manipulating life processes, which underpinned this research, fit well with the Bolsheviks' general technocratic and utilitarian attitudes toward science and its place in the new social order they were creating.[22] What's more, this ideal resonated strongly with what Richard Stites (1989) has fittingly called "revolutionary dreams" of creating a "new society" and a "new world," which found particularly fertile ground and enjoyed enormous popularity in postrevolutionary Russia.[23]

The Bolshevik Revolution launched the world's first "great experiment": the realization of these centuries-old visions on the one-sixth of the world under the Bolsheviks' control.[24] Inspired by these "revolutionary dreams," the nation experimented on an unprecedented scale with new ideas and practices in every facet of its life, ranging from the arts to industry, from state administration to education, from laws to technology, and from literary works to community services. Not surprisingly, the ultimate experimentalists—scientists—were also captivated by these "dreams."

Furthermore, the broader social and cultural implications of the budding union between "visionary biologists" and "revolutionary dreamers" attracted the close attention of the contemporary literary community. As elsewhere in the world, a new literary genre—science fiction (*nauchnaia fantastika*)—flourished in 1920s Russia, filling the pages of daily newspapers, weekly magazines, popular-science journals, and books.[25] As one of the pioneers of Soviet science fiction, Evgenii Zamiatin, put it in his 1922 essay about the acknowledged founder of the genre, H. G. Wells, "Russia, which during the last few years has become the most fantastic country of modern Europe, will undoubtedly reflect this period in its fantastic literature" (Zamiatin 1922, 47). Zamiatin's prediction quickly came true.

Science fiction took Soviet Russia by storm, and a large subset of this new literature found its inspiration in experimental biology and medicine. Dozens of novels, poems, and innumerable short stories projected future applications of the contemporary advances in various fields of experimental biology and medicine, exploring their possible impact on individuals, societies, and the world at large.[26] These works assessed the role of science and scientists, state agents and agencies, technologies and personalities, in bringing about a "new society" and a "new world." Some of them were authored by scientists and physicians who themselves pursued exciting new lines of research; others by professional fiction writers and journalists. Some were extremely enthusiastic, hailing a bright future being ushered in by the power of science. Others were deeply pessimistic, arguing that instead of creating a "new world," science would amplify the world's misery. But practically all of them enjoyed enormous popularity with readers.

Nowhere did this "conversation" among revolutionary dreamers, visionary biologists, and science-fiction writers—albeit, in this case, mostly an "internal" one—reach such intensity and expression as in the life, writings, and work of Alexander Bogdanov. His revolutionary dream of a new socialist society included from the start a major biological vision of humanity "physiologically united" and "rejuvenated" by blood exchanges. He was one of the first to explore—in his fiction and philosophical treatises—the possible implications of the interactions between science and society, as well as between social revolution and biological evolution. He also was one of the first to attempt, not in fiction but in a research institute, to investigate exactly how his vision could be realized. And last but not least, as one of the founding members of the Bolshevik Party, Bogdanov exerted considerable influence on its leadership, even though he had left the party long before—and refused to rejoin it after—the Bolsheviks came to power in Russia.

An examination of Bogdanov's works therefore provides a unique window into the interplay between the Bolshevik Revolution and the experimental revolution in the life sciences in its institutional, intellectual, cultural, and social dimensions. An analysis of Bogdanov's three personas—a revolutionary dreamer, an SF writer, and a visionary biologist—presents an unparalleled opportunity to explore interactions between science and society, ideology and institutional development, scientific ideas and societal values, in revolutionary Russia. It also allows us to examine the processes of transnational borrowing, translation, and "acclimatization" of particular scientific ideas, techniques, and institutions and, at the same time, brings into sharp relief the instances and origins of true innovations. I believe that such an analysis allows us to place the Russian/Soviet case on the world's historical maps with greater precision.

SCIENCE: MARXIST AND PROLETARIAN

In particular, such an examination offers a novel perspective on the long-contested subject of the interrelations between the life sciences and Marxism in Soviet Russia. Ever since the Bolshevik Revolution, but especially with the beginning of the Cold War, both participants in and observers of Soviet science have hotly debated the nature of these interrelations. The debates have produced three distinct sets of arguments (and correspondingly, three subsets of literature), complete with matching examples drawn from the history of Soviet life sciences.

Many scholars have forcefully argued that as a philosophical doctrine, Marxism influenced both the intellectual and institutional development of Soviet science. On one hand, they have documented the extraordinary growth of disciplines, personnel, and institutions under the new regime as the reflection of a Marxist belief in the power and utility of science. On the other hand, they have examined works of individual scientists (ranging from the biochemist Alexander Oparin to the geneticist Alexander Serebrovskii, the morphologist Ivan Shmalgauzen, and the physiologist Ivan Pavlov) whose research and writings manifest convincingly the important role that Marxist dialectical materialism played in the highly original ideas developed by these scientists and their students.[27] Surprisingly, although this subset of literature often invokes Bogdanov as an influential Marxist philosopher, neither his concept of science nor his own research on blood transfusion has commanded much attention.

Other scholars have no less convincingly demonstrated that many Soviet scientists adopted and adapted the official "Marxist" lexicon to pursue their own institutional, intellectual, and personal ambitions under the new regime, even though concerted efforts by Marxist philosophers and ideologues during the 1920s to "infuse" Marxism into the actual practice of Soviet science had largely failed.[28] This subset of literature often discusses Bogdanov's writings in some detail as one among the multitude of diverse and divergent trends in Soviet Marxism, which exerted considerable influence on its leading proponents (from Vladimir Lenin to Anatolii Lunacharskii to Nikolai Bukharin), its internal debates, and its subsequent developments in Soviet Russia. But this literature also barely touches upon Bogdanov's concept of science or his own exploits in the science of blood.

Still other scholars have maintained that Marxism deeply—and largely negatively—affected the development of Soviet science and in fact resulted by the end of the 1920s in the emergence of a particular "proletarian science" profoundly different from science elsewhere. The favorite cited example of this development is the notorious Michurinist "agrobiology" propagated by Trofim Lysenko.[29] This subset of literature points an accusing finger at Bogdanov as the theoretical "father" of "proletarian" biology. Indeed, one of the leading proponents of this view, philosopher Dominique Lecourt, proved instrumental in bringing out a French edition of Bogdanov's collected works on the subject of proletarian science (Bogdanov 1977) and even provided it with his own introduction, eloquently entitled "Bogdanov, the mirror of the Soviet intelligentsia." Still, this literature thoroughly ignores Bogdanov the scientist.[30]

Needless to say, these three subsets of literature rarely engage in a fruitful conversation, even though their central arguments are hardly mutually exclusive.[31] I believe that a careful examination of Bogdanov's life, writings, and research could contribute substantially to this long-standing debate. Bogdanov was one of the very small number of individuals among the first generation of avowed Russian Marxists who was directly involved both in articulating what science in a socialist society ought to be and in actually practicing science in Soviet Russia.[32] His science therefore provides a unique case study for a pointed exploration of the interrelations between science and Marxism, in all of their rich intellectual, institutional, and cultural dimensions.

I begin with a brief outline of the history of blood transfusions in Russia prior to the establishment of Bogdanov's institute, discussing the discipline-building efforts of scientists and their patrons in Narkomzdrav during the first years of the Soviet regime. I examine the ways information about blood transfusions filtered into Russia and explore the particular dynamics of relationship between scientists and their patrons, which hampered the institutional development of "blood science." I then trace the evolution of Bogdanov's dream of a future socialist society and his vision of humanity united by blood exchanges from *Red Star* to *Tectology*. I examine the sources of his vision and its role in his later works, bringing together Bogdanov's three personas as a revolutionary, a writer, and a scientist through a detailed analysis of his concept of "proletarian science" and its place in his philosophical and fictional writings. I proceed to describe the circumstances of Bogdanov's move from theoretical speculation to actual experiments with blood exchanges in 1924–25, focusing on specific medical contexts of the mid-1920s, particularly on the epidemic of "Soviet exhaustion," which plagued the country's ruling elites. I argue that concerns and anxieties this epidemic generated among the Bolshevik leadership, including Politburo members Josef Stalin, Aleksei Rykov, and Leon Trotsky and commissars Leonid Krasin and Nikolai Semashko, overrode the opposition of medical officials to the establishment of a special institute for blood research and prompted Bogdanov's appointment as its director. I then assess the institute's activities during Bogdanov's tenure from the spring of 1926 to the fateful experiment that led to his death in April 1928. I look particularly at Bogdanov's 1927 manifesto *Struggle for Viability,* which applied the basic principles of his "proletarian science" to the studies of blood transfusion and articulated a "tectological" theory of senescence and rejuvenation as the theoretical foundation of both his vision of "physiological collectivism" and his research program on blood exchanges. I examine the "death" of Bogdanov's vision following his own death and the subsequent transformation of his institute into the fulcrum of a countrywide system of blood services that emerged in the Soviet Union during the early 1930s, spurred by Stalin's "revolution from above" and the ensuing extensive militarization of the country. I conclude with an exploration of multiple legacies Bogdanov bequeathed to Soviet science, particularly his concept of "proletarian science," which, despite the vicious public condemnation of its author in Stalin's Russia, provided a foundation for much of Soviet science policy for years to come.

2 Transfusions

IDEAS, PEOPLE, AND PLACES

For most of its history, science in Russia developed in close contact with and under the profound influence of its counterparts abroad. Indeed, until the mid-nineteenth century, the majority of Russia's premier scientists were foreigners, and after that, nearly all of Russia's leading scientists spent some time abroad studying, doing postgraduate work, attending conferences, and, occasionally, lecturing at various scientific institutions in Western Europe. Not surprisingly, until the Bolshevik Revolution of 1917, Russian developments in experimental biology and medicine in general—and in blood transfusions in particular—closely paralleled those in other European countries.[1]

Following British obstetrician James Blundell's pioneering and widely publicized attempts at blood transfusions in the mid-1820s,[2] Russian physicians occasionally experimented with the procedure, as did their colleagues elsewhere.[3] During the 1830s and 1840s, several doctors at the Medical-Surgical Academy (later renamed Military-Medical Academy) in the country's capital, St. Petersburg, used blood transfusions in their own gynecological practices.[4] They published accounts of their experiences in the academy's *Military-Medical Journal,* reported them to professional conferences, and even included sections on blood transfusions in their surgical manuals. In 1848, Moscow University's professor of physiology Aleksei Filomafitskii (1848) published a three-hundred-page *Treatise on Blood Transfusion (as the Only Means in Many Cases to Save a Fading Life), Compiled in Historical, Physiological, and Surgical Perspectives.* The treatise summarized existing literature, analyzed contemporary theoretical views on and practical methods of blood transfusions, and described Filomafitskii's own animal experiments with transfusions of "blood serum" (what today would be called "blood plasma"). Filomafitskii's students extended his experiments to clinical practice. In 1865, Vasilii Sutugin, an assistant at the gynecological clinic of the Medical-Surgical Academy, defended the first doctoral dissertation "On Blood Transfusion" in Russia. The dissertation presented results of Sutugin's experiments on dogs with transfusions of blood kept in cold storage for up to seven days, comparing the effects of transfusing "whole blood" and "blood serum" (Sutugin 1865). In the 1870s, physicians in St. Petersburg, Moscow, Kazan, and other university centers continued animal experimentation and occasionally performed transfusions of both human and animal blood to their patients. But as elsewhere in the world, the uncertainty of results prevented widespread use of the technique in Russia: by the 1880s, various saline solutions had replaced blood both in clinical practice and in physiological experiments.[5]

For the next quarter century, blood transfusions in Russia were almost completely abandoned. As Fedor Borngaupt, a well-known professor of surgery at Kiev University, put it jokingly in 1903: "To perform a blood transfusion you need three lambs: one—to whom blood is transfused; another—from whom it is transfused; and the third—who transfuses it."[6] As were many of their colleagues in Europe—particularly in their favorite destinations for postgraduate studies, Germany and France—Russian physicians were slow to recognize the importance of the discovery of blood groups for transfusions.[7] Although, during the Great War, Russia fought on the side of the Allies, Russian military surgeons did not practice blood transfusions as did their British and French counterparts, since no Canadian or U.S. doctors joined the Russian army to transfer their knowledge of the procedure to their Russian colleagues as they had with the French and British.[8] This North American know-how, along with the Great War experiences, eventually filtered into Russia, but this happened largely after the Bolshevik Revolution of 1917, which created entirely new ideological, political, and institutional landscapes for the country's medical services and biomedical research.

THE BOLSHEVIK REVOLUTION AND
EXPERIMENTAL BIOMEDICINE

On October 25, 1917, a radical faction of the Russian Social-Democratic Labor Party, the Bolsheviks, carried out a coup d'état in Petrograd[9] and declared the establishment of a socialist republic. In early March 1918, the Bolsheviks concluded a separate peace treaty with Germany, ending Russia's participation in World War I. But within a few weeks, a civil war erupted between the "Reds" and the "Whites," with various local "Green" groups fighting both across the country's vast expanses.[10]

Threatened by the White forces advancing on Petrograd, the Bolshevik leadership moved to Moscow, which became the capital of the new, Soviet Russia. With the civil war escalating, in the summer, the Bolsheviks adopted an economic policy of "war communism," which featured the nationalization of all enterprises, the abolition of private property and money, the forced requisition of agricultural production, and the administrative distribution of food and goods. In Moscow, the Bolsheviks began to create a new government apparatus: on July 10, 1918, a Congress of soldier, peasant, and worker deputies adopted the first constitution of the Russian Soviet Federated Socialist Republic (RSFSR).

The next day, July 11, the leader of the Bolsheviks, Vladimir Lenin, signed a decree establishing a special agency to "protect the health of the people" in the newborn republic: the People's Commissariat of Health Protection (Narkomzdrav).[11] Issued by the highest government body, the Council of People's Commissars (SNK), the decree appointed Nikolai Semashko, a Bolshevik doctor, to head the new commissariat.[12] The new agency immediately began setting up regional offices and a central apparatus to fulfill its mandate. A month later, Narkomzdrav instituted a Scientific Medical Council—a consultative body comprised of eminent medical researchers and practitioners chaired by Lev Tarasevich, the country's leading bacteriologist (and a longtime Bolshevik sympathizer)—to tap the expertise of Russian physicians in creating a new system of medical services and biomedical research in the country. In the early fall of 1918, Narkomzdrav established a State Institute of People's Health Protection (GINZ), an association of research institutions that was to address questions pertaining to the entire field of medicine and public health.[13] In October, following the recommendation of the Scientific Medical Council, Narkomzdrav approved the founding of the first research institute that became a part of GINZ—the Institute of Tropical Diseases—to lead the fight against a growing epidemic of malaria in the country. By the end of 1921, GINZ had acquired seven more institutes for research in microbiology, social hygiene, vaccines and sera, biochemistry, tuberculosis, experimental biology, and the physiology of nutrition.[14]

In late 1921, the civil war ended. Although the fight against various "Green" groups continued in different regions of the country for several more years, the Red Army had driven out both the "Whites" and the Allied expeditionary forces that supported them. But the Bolsheviks paid a steep price for the victory: seven years of continuous warfare had left the country on the brink of collapse. Faced with imminent economic crisis exacerbated by famine, epidemics, and civic unrest, the Bolshevik government abandoned war communism and inaugurated a "new

17

economic policy"—the NEP. Preserving state control over key industries and banking, the Bolsheviks partially restored private property and private initiative in trade, services, and the manufacture of consumer goods. They abolished the forced requisitions of agricultural products and reinstated money. The NEP proved highly effective in revitalizing the country's economy: in just two years, Russia's industrial and agricultural production were revived.[15]

The NEP also greatly facilitated the institution-building efforts of Russian scientists and their patrons in the government apparatus. In the early 1920s, in response to Russian scientists' continuous lobbying for building "big science" institutions, Narkomzdrav generously funded the organization of large research institutes in a variety of biomedical fields. New disciplines, ranging from biochemistry and biophysics to genetics and endocrinology, were quickly institutionalized under Narkomzdrav's auspices. Not surprisingly, a number of "the world's first" specialized research institutions were created in Russia during this time. By the spring of 1926, when Semashko's order created the Institute of Blood Transfusion, Narkomzdrav had already founded nearly forty specialized research institutions, ranging from the Institute of Social Hygiene and the Institute for the Control of Vaccines and Sera to the Institute of Occupational Diseases, the Institute of Experimental Endocrinology, the Institute of Biophysics, and the Institute for the Protection of Maternity and Infancy, with the majority of these new research establishments located in Moscow.[16]

The first puzzle, then, is not why the "world's first" Institute of Blood Transfusion was organized in the Soviet Union in 1926 but why it was founded so late. One would expect that in the frenzy of institution building that seized the newborn Soviet state during the postrevolutionary era, such an institute would have been established under Narkomzdrav much earlier. But, until the mid-1920s, the institutional development of blood research lagged behind other fields of biomedicine as a result of the particular dynamics of interactions between Russian researchers and their patrons in the fledgling government apparatus.

An examination of Narkomzdrav's archives reveals that several physicians did lobby the agency for creating special institutes to study blood and blood transfusions well before 1926.[17] Indeed, as early as November 1920, while the civil war was still raging throughout the country, the Scientific Medical Council received a proposal from Savelii Tsypkin, a well-known Moscow internist, to establish a "hematological institute for the treatment and study of blood diseases."[18] At its meeting on December 3, 1920, the council discussed the proposal.[19] Its members decided that a "study of the pathogenesis of various blood diseases from pathological-anatomical, biochemical, and serological points of view could be successfully carried out at existing clinical establishments, as well as at existing institutes of biochemistry and serology," and concluded that creation of a separate hematological institute was unnecessary.[20]

Perhaps the shortages and deprivations of the civil war and war communism could account for this decision. Yet these factors could not explain a similar decision adopted by the council some two years later: the NEP was in full swing, and by then Narkomzdrav had far more resources to spend on building research institutes. In June 1923, Nikolai Kukoverov, a professor of surgery at the Odessa Medical School, proposed the establishment of an institute for research on blood and blood transfusions. He presented a long memorandum on the issue to the Narkomzdrav Main Military-Sanitary Directorate.[21] After outlining the early history of blood transfusions and the discovery of blood groups, Kukoverov detailed the experiences with blood transfusions gathered during the Great War, which had prompted important developments in the technique and the organization of services for blood transfusion in the West, particularly the United States. He even provided a brief summary of reports on blood transfusions presented at the first postwar international congress of surgeons held in 1920 in Paris. He also described the efforts of Odessa surgeons and researchers—endorsed by the Odessa Medical Society—in developing reliable transfusion techniques. In conclusion, Kukoverov urged the Main Military-Sanitary Directorate to consider the subject of blood transfusions a priority for the Red Army's medical services and suggested that a special institute "to study the properties of blood in relation to transfusion" be established.

The directorate forwarded Kukoverov's memorandum to the Scientific Medical Council for expert evaluation. To assess the merits of the proposal, the council set up a special commission that included the council's chairman Tarasevich; Nikolai Burdenko, a rising star surgeon; and Petr Diatropov, a prominent microbiologist and the dean of Moscow University's medical school. As it happened, Burdenko—the only member with any experience in transfusions—wrote the commission's "assessment" for the council. Burdenko's experiences with the operation, however, were quite limited: he himself had performed only two transfusions, and both had been unsuccessful. Burdenko underestimated the importance of transfusions for medicine in general and military medicine in particular. He claimed that even during the Great War, the need for blood transfusions had been very low (he estimated that fewer than one thousand transfusions had been performed by all the armies involved in the war) and it could not be much higher

19

in the future. The council accepted Burdenko's "verdict" and rejected Kukoverov's proposal.

Yet even though the Scientific Medical Council rejected the idea of organizing a special institute for blood research at least twice, such research was already being conducted in existing medical institutions throughout the country. Indeed, as early as June 1919, Vladimir Shamov had performed the country's first reported blood transfusion matching the blood groups of the donor and the recipient at a clinic of the Military-Medical Academy (VMA).

THE PIONEER

Vladimir Shamov (1882–1962) was born into the family of a schoolteacher.[22] In 1901, after graduating from a classical gymnasium in Perm', near the Urals, he enrolled into the VMA in St. Petersburg. He was a diligent student but took an active part in student strikes and street protests during the first Russian revolution of 1905–6, as did many of his fellow students. As a result, he was arrested, imprisoned, and then exiled from St. Petersburg for one year. Shamov was able to resume his studies, and in November 1908, he graduated from the academy with first-class honors. VMA professors marked Shamov as a gifted student, and, unlike his classmates who entered military service and were stationed in remote garrisons scattered throughout the empire, Shamov stayed in St. Petersburg, chosen "to prepare for the title of professor"—a Russian analogue of what we would today call graduate studies. For the next three years, he worked at the VMA surgical clinic under the tutelage of the clinic's director, Sergei Fedorov (1869–1936), a prominent military surgeon.[23] After the successful defense, on January 28, 1912, of Shamov's doctoral dissertation "On the Significance of Physical Methods for the Surgery of Malignancies" (Shamov 1911), Fedorov appointed the newly minted "doctor of medicine" to his clinic as assistant surgeon.

In early 1913, Fedorov became the chief surgeon of the Russian army, and that summer he sent his assistant abroad to study systems of military-medical education and learn the newest advances in surgical techniques. It is unclear from available sources whether Shamov himself or his patron Fedorov chose the destination of his travels. Whatever the case, unlike the majority of Russian physicians who did their postgraduate studies in Germany and occasionally France, Shamov went to England for three months and then to the United States for almost nine months. He visited medical schools and hospitals everywhere he went and sent home detailed reports on his findings and experiences (Shamov 1914a).

2.1 Vladimir Shamov after his graduation from VMA, circa 1909.
(Courtesy of the Military Medicine Museum, St. Petersburg.)

The American part of his trip proved particularly exciting and reward-ing: Shamov met all the leading lights of American medicine of the time. He worked with William James Mayo in Rochester, Minnesota; with Alexis Carrel at the Rockefeller Institute for Medical Research in New York City; with Harvey Cushing at Harvard Medical School in Boston; and with George Washington Crile at Case Western Reserve School of Medicine in Cleveland, Ohio (Shamov 1914b, 1915). He carefully stud-ied the organization of the Mayo clinics. He closely observed Carrel's famous experiments on tissue cultivation and organ transplants. He spent several months at Cushing's laboratory in Boston, conducting experiments on the physiology of the pituitary gland and the vagus nerve and—with Cushing's assistance—eventually publishing his results in the *American Journal of Physiology* (Boothby and Shamov 1915; Shamov 1915–16a, 1915–16b). But it was at Crile's clinic that Shamov first witnessed successful blood transfusions from human to human and learned the techniques of matching the blood groups of donor and recipient.[24]

At the time, the United States led the way in research on blood transfusions, and Crile was arguably the country's leading proponent of introducing the operation into clinical practice.[25] In the early 1900s, inspired by Carrel's technique of artificial anastomosis—the surgical connection of blood vessels—Crile attempted to use this technique for blood transfusions.[26] After several dozen experiments on dogs, in 1906 he successfully performed a direct blood transfusion from human to human, surgically connecting the donor's artery with the recipient's vein. Crile (1906a, 1906b, 1907) reported his observations at a meeting of the Society of Experimental Biology and Medicine and published them in leading U.S. journals: *Science, Annals of Surgery,* and *JAMA.* Carrel's technique of suturing blood vessels, however, was extremely delicate, difficult, and time-consuming. It involved stitching together very thin blood vessels (three to five millimeters in diameter), using the tiniest needles and finest silk threads, in such a way that the seam would not leak any blood and at the same time would not prompt the formation of blood clots and thus obstruct blood flow through the anastomosis. Crile simplified Carrel's technique by designing a special silver canula to connect the two vessels, which made the whole operation much easier and faster, and thus allowed Crile to greatly expand the use of blood transfusions both in experiments and in the clinic. In 1909, Crile published a 560-page treatise entitled *Hemorrhage and Transfusion,* which described in great detail—with more than sixty illustrations—both Carrel's and his own techniques of artificial anastomosis, the requirements for the operating room and the preparation of donors and recipients, and the results of his numerous experimental and clinical studies on the use of blood transfusions (Crile 1909). By the time Shamov came to Cleveland, Crile had also adopted methods of cross-matching blood of the donor and the recipient based on Karl Landsteiner's discovery of different human blood groups and William L. Moss's refinements of Landsteiner's findings.[27]

Conditioned by Russian (and more generally European) physicians' disdain for blood transfusions, Shamov was definitely not prepared for his experiences at Crile's clinic. As he himself readily admitted, "Having come to America, I perceived with deep skepticism the reports that the

Americans had again begun to perform blood transfusions, the operation that the European science seemed to have proved unequivocally to be unscientific and useless." Shamov was "astonished" to see that in Crile's clinic, "blood transfusions not only are not accompanied by any complications and dangers, but to the contrary they produce very demonstrable, simply 'miraculous' results." The scientist in Shamov prevailed. As he later recounted, "Having observed the results obtained by Crile, I quickly changed from a deep skeptic into an enthusiast and convinced proponent of the method of blood transfusion" (Shamov 1937, 787).

The onset of World War I in the summer of 1914 cut Shamov's trip short. In late August, he returned to St. Petersburg to resume his appointment as a VMA assistant surgeon. He kept working at Fedorov's clinic throughout the war, the Russian revolutions of 1917, and the civil war. The trip abroad made an indelible impression on the young surgeon. As he intimated in a letter to Cushing written in March 1915, "Now that I have my Boston time behind me, I can really appreciate it, and I wanted to tell you with an open heart that I shall never forget what I have learned from you."[28] Yet the physiological experiments he had conducted and new surgical techniques he had observed in Cushing's laboratory were not the only things he had learned abroad. Ever since his return from the United States, Shamov dreamed about employing his newfound expertise in blood transfusions in his own clinical practice. But the reality of Russian life cooled his enthusiasm. To begin with, Shamov's effusive tales of Crile's successes with blood transfusions failed to convince his VMA colleagues, who remained highly skeptical regarding this "discredited" operation. There was one exception, however—Shamov's patron, Sergei Fedorov, who had always trusted his star student's judgment and allowed him complete freedom in his ward. But another, much more serious and quite unexpected obstacle thwarted Shamov's efforts: he could not find a donor. "When I raised the issue of transfusion of others' blood, patients and their relatives did not want to even hear about this 'horrifying experiment.' Despite all my persuasions, not a single person was willing to give blood for transfusion" (Shamov 1937, 786). Adding to his troubles, the war made it impossible for Shamov to get the standard blood sera for determining blood groups of the donor and the recipient from the United States. Still, he persevered.

Finally, in June 1919—at the height of the civil war—Shamov found an opportunity to realize his dream and to perform blood transfusions at the VMA clinic. In his recollections, written some twenty years after the fact, he provided a poetic account of the first attempt (Shamov

1937). Shamov had a patient with a large cervical tumor in the clinic's gynecological ward. He wanted to excise the tumor but feared that the patient was too weak and anemic to survive the operation because of repeated excessive bleeding from the tumor. Shamov decided to try a blood transfusion. Since no standard sera to determine the patient's blood group and to find a matching donor were available in Petrograd at the time, Shamov set out to create his own standards. With the help of several assistants, including a graduate student, Nikolai Elanskii, Shamov collected blood samples from twenty-two of the academy's workers and students. He cross-matched all the samples and was able to find a suitable donor—a young woman who worked at the academy as a typist. It took Shamov several days (and the promise of a paid vacation) to convince the woman that giving blood for transfusion would not do her any harm but rather could save the life of his patient. The woman finally agreed. Yet there remained one more obstacle: Shamov could use neither Carrel's nor Crile's technique of direct transfusion, which he had personally observed in the United States. He did not have enough experience in Carrel's method of suturing the blood vessels and lacked the necessary tiny needles and silk threads. Nor did he have Crile's canulae. So Shamov decided to use the much simpler method of indirect transfusion with citrated blood, which had become popular during World War I among the Allied surgeons.

Of course, despite his firm belief in the mighty science behind Western surgeons' successes with blood transfusions, Shamov felt quite uncertain about the outcome of his own first try. The skepticism of his colleagues was infectious. A number of "what ifs" revolved in his mind:

> What if the favorable results he had observed in the United States were just a coincidence? What if the issue of the compatibility of donor's and recipient's blood is not limited only to determining their blood groups? What if there are other, yet unknown factors involved? What if precisely in his own case these unknown factors led to the death of his patient from the transfusion of another's blood, as had happened more than once in the past practice of the operation? What if the young woman who so trustingly decided to give her blood for the transfusion got sick as a result? What if she developed tuberculosis? (Shamov 1937, 787)

With shaking hands, he started the procedure under the inquisitive gazes of his skeptical colleagues. In the end, science triumphed: the collection and then the transfusion of 570 milliliters of citrated blood went smoothly, with no complications. The donor "easily withstood the bloodletting." The patient's condition improved dramatically, and several days later, Shamov was able to excise the tumor, which turned out quite successfully—the patient recovered and was discharged from the clinic.

Of course, in his recollections, Shamov somewhat embellished his early experience with blood transfusions: his actual first attempt at using the procedure had been much less dramatic and "miraculous." In fact, Shamov's first case was an eleven-year-old boy with a large tumor at the base of his skull. To "fortify" the boy before the surgical removal of the tumor, Shamov transfused him with 100 milliliters of blood from his mother. A blood test performed after the transfusion confirmed that the boy's blood formula had improved: the number of red blood cells (erythrocytes) grew from 3,750,000 to 4,200,000, and the hemoglobin level increased from 60 percent to 70 percent. Shamov decided that he would undertake the planned excision, but due to extensive loss of blood during the operation, the patient died, despite all Shamov's efforts to revive him.

Obviously, this case did not discourage Shamov from continuing his explorations with blood transfusions, and the second case, which he so fondly recounted in his recollections, strengthened his determination. His third case also involved a young woman operated on in his ward. Shamov had removed a large tumor from her pelvic bone, but the patient went into postoperative shock. Shamov transfused her with 420 milliliters of blood from her brother, and the woman recovered: her blood pressure and blood formula went back to normal.

Inspired by his success, in June 1920, Shamov presented a long report on blood transfusions at a general conference of the VMA staff (Shamov 1921). He began with a brief excursion into the history of the operation, describing the deep disillusionment about its clinical value that had emerged toward the end of the nineteenth century and "remained among the majority [of physicians] up to this day." But, Shamov argued forcefully, "this situation must change." Although in his presentation Shamov referred directly to only one source, Crile's 1909 book, his report demonstrated that he kept abreast of available literature on the subject. Thanks to Crile's pioneering work, Shamov stressed, "The method of blood transfusion had begun to develop quickly . . . and has already found wide clinical application in North America. In the last few years, it has also begun to be practiced in France and England, and infrequently in Germany." Shamov emphasized that Crile's success in the application of this once-discredited technique rested on solid scientific and clinical foundations: the discovery of blood groups and the development of surgical techniques for artificial anastomosis. He described in detail the principles and concrete methods of determining blood groups, as well as Carrel's and Crile's techniques, noting that subsequent developments had led to their further simplification. He also stated that during the last few years, with the discovery of sodium citrate's anticoagulation properties, the method of indirect transfusion began to replace the surgically demanding methods of direct transfusion.

Shamov informed the audience of his own success in creating the standard sera for determining blood groups and briefly described his three clinical cases, displaying several tables that showed the results of his patients' blood tests before and after the transfusions. He concluded by outlining possible areas of clinical application for blood transfusions and highlighting "all the value of this method for surgery, as well as internal medicine." Still, staff members of other VMA clinics, who had not witnessed Shamov's actual transfusions, greeted his report with the customary skepticism and disbelief.

Not discouraged by his VMA colleagues' cool reception of his presentation, Shamov remained determined to popularize blood transfusions among Russian surgeons and to extend the use of this "lifesaving" technique in his clinic. As always, his patron extended a helping hand. With the civil war subsiding, in early 1921 Fedorov managed to establish a new journal, the *New Surgical Archive*. The editor in chief certainly used his editorial powers to include the text of Shamov's report to the VMA conference in the first issue of his journal (Shamov 1921). In September, Fedorov took a leave of absence from his VMA clinic and appointed Shamov as its acting head. With Fedorov's blessings, Shamov immediately organized a small group of undergraduate and graduate students to study blood groups, to manufacture the standard sera for blood typing, and to expand the use of blood transfusions in his own clinical practice.

Shamov's 1921 article—the first original publication on the subject of blood transfusions in Soviet Russia—resulted in a stream of letters from physicians all over the country asking for details of Shamov's operations, particularly the determination of blood groups, as well as sera for blood typing. Shamov's group generously sent out their standard sera along with detailed instructions on how to use them. But serving as a supplier of sera for Russian physicians was obviously not something Shamov's busy surgical clinic could afford to do on a regular basis. There had to be other ways of popularizing blood transfusions. In May 1923, Shamov and Elanskii delivered a joint report to the Pirogov Society of Russian Surgeons in Petrograd, and a few months later, they published it in the *New Surgical Archive* as a thirty-page article. Entitled "Isoagglutination Properties of Human Blood, Their Importance for Surgery, and Methods of Their Identification," the article surveyed more than

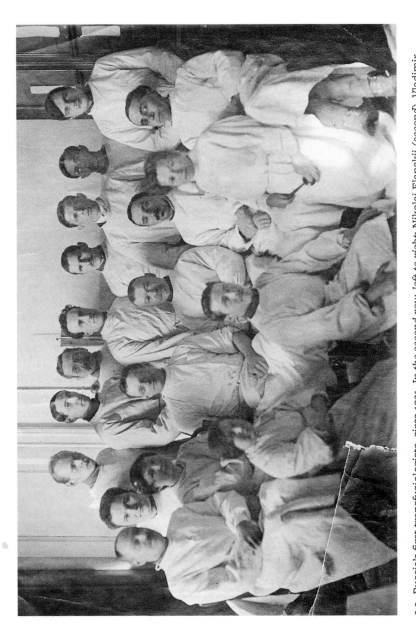

2.2 Russia's first transfusiologists, circa 1921. *In the second row, left to right*: Nikolai Elanskii (*second*), Vladimir Shamov (*fourth*), and Sergei Fedorov (*fifth*). (Courtesy of the Military Medicine Museum, St. Petersburg.)

sixty foreign publications on the subject of blood groups, detailing available techniques of blood-typing, along with methods of producing standard sera (Shamov and Elanskii 1923).

In the fall of the same year, Shamov left Petrograd to accept the position of head surgeon at the Kharkov University Medical School. The group he had created at the VMA surgical clinic, with the active support of its returned head, Fedorov, continued his work on blood transfusions. Shamov's former assistant Nikolai Elanskii (1894–1964) assumed a leadership role. The following year, in a series of "university manuals" he was editing, Fedorov issued the Russian translation of a short book on blood transfusions written by French surgeon Dupuy de Frenelle (1923). In his introduction to the book, Fedorov stressed the importance of blood transfusions and recommended that all doctors incorporate this "lifesaving" operation into their clinical repertoire (Frenel' 1924). As a reviewer in *Moscow Medical Journal* emphasized, the book indeed described "the technique of transfusions that could be used by anyone."[29] In early 1925, a leading medical journal, *Russian Clinics,* carried a long article on the causes of various complications observed after blood transfusions, written by Fara Korganova-Miuller (1925), a member of the VMA transfusion group. And at the beginning of the following year, Elanskii published a two-hundred-page monograph on blood transfusions, issued as part of "the library of a practicing physician," also edited by Fedorov. Prefaced by Fedorov's enthusiastic praise, the book surveyed available literature, recounted the methods of blood typing, described the techniques of direct and indirect transfusions, and mapped out possible fields of application, thus providing "practicing physicians" with all the knowledge necessary for using blood transfusions in their daily practice (Elanskii 1926).

THE ENTHUSIASTS

Shamov's pioneering publications stimulated a number of Russian surgeons and internists to take the issue of blood transfusions seriously and start practicing the procedure.[30] From 1923 to 1926, they performed transfusions in Leningrad and Moscow, Tashkent and Odessa, Kharkov and Tomsk, Baku and Voronezh, Kazan and Omsk, Saratov and Kiev.[31] To give but one example, on May 6, 1923, Valentin Voino-Iasenetskii, a prominent surgeon, reported to a meeting of the Turkestan Medical-Scientific Society in Tashkent on the blood transfusions he had performed on anemic patients.[32]

Of course, given the elaborate network of professional institutions, societies, meetings, and publications in Leningrad, the VMA group's experiences found the most receptive audience in the city's medical establishments, particularly at the Institute for Doctors' Continuing Education. On February 2, 1925, Erik Gesse (1883–1938),[33] the chairman of the institute's surgery department, delivered a special lecture on the subject, encouraging doctors and students to learn the techniques of blood typing and blood transfusions (Gesse 1925). Ten days later, Iosif Maiants (1925), one of Gesse's assistants, presented a long report on "blood transfusions in gynecological and surgical practice" to a meeting of the Leningrad Society of Gynecologists and Obstetricians attended by nearly one hundred members. Blood transfusions attracted the close attention of gynecologists and obstetricians, who more than any other disciplinary group among physicians dealt with excessive blood loss in their daily practice. The same year, Alexander Mandel'shtam (1925), an assistant professor at the gynecological clinic of the Institute for Doctors' Continuing Education, published a description of a simplified method of blood typing he had developed. In May 1926, another member of the clinic presented a report on blood transfusions to the All-Union Congress of Gynecologists and Obstetricians (Barskii 1927). The next year, staff members of the clinic put together a manual on "blood transfusion for doctors and students," with an extensive bibliography of more than one hundred publications in Russian and foreign languages available in Leningrad libraries (Krinitskii and Rutkovskii 1927). Gesse (1926b) also published an extensive survey on "indications for blood transfusions" in one of the most widely circulated professional periodicals, *Physician's Gazette*.

In the capital, several researchers and clinicians at Moscow University's renowned medical school took up studies in blood transfusions. As early as 1923, two assistants at the university clinics conducted a series of experiments on rabbits to determine the fate of transfused red blood cells (erythrocytes) in the recipients' blood (Shustrov and Dmitriev 1923). One of the most active proponents of the new operation was Iakov Bruskin (1888–1972?), a lecturer at the university's medical school and a surgeon at its Oncology Institute.[34] In 1923, thanks to the close alliance of the two "pariahs" of Versailles—Weimar Germany and Soviet Russia—which broke the nearly total international isolation of the newborn socialist republic in general, and its scientists in particular, Bruskin was able to spend six months in Germany visiting various hospitals and clinics to learn the newest advances of German surgery. In one hospital in Hamburg, he observed direct blood transfusions performed by local surgeons and, by his own admission, became a convert (Bruskin 1924).

At the beginning of 1925, Bruskin published a long survey on the use of blood transfusions in a leading journal, *Herald of Modern Medicine*.

Unlike the Leningrad group, Bruskin considered the transfusion of citrated blood inefficient, if not downright harmful, and strongly advocated transfusions of whole blood directly from donor to recipient (Bruskin 1925b). On February 23, 1925, the Russian Surgical Society in Moscow held a special session on blood transfusions with two keynote speakers, Bruskin and Korganova-Miuller.[35] Understandably, the presentations by two leading experts in the field spurred a lively discussion at the meeting. Two days later, on February 25, Bruskin also delivered a report on "the newest methods of blood transfusion and their achievements in the field of therapy" to the Moscow Therapeutic Society. Notes on and the transcripts of these two meetings appeared in almost every medical journal in the country.[36] He also published an article on the technique of blood transfusions in Narkomzdrav's official journal, *Medical Worker* (Bruskin 1925a). But Bruskin did not limit his propaganda efforts to medical audiences. In January 1926, a popular daily, *Evening Moscow,* published a long article based on Bruskin's interview, detailing various techniques of blood transfusion and medical conditions that required it ("Novyi sposob . . ." 1926). At the time Narkomzdrav issued the order creating Bogdanov's institute, Bruskin (1926) also published an article on blood transfusions in a popular Communist magazine, *Young Guard.* On March 1, 1926, Bruskin gave another talk at a meeting of the Russian Surgical Society and demonstrated a new apparatus he had designed for direct blood transfusions.[37] A week later he again addressed the society, this time presenting results of his experiments on dogs aimed at the elucidation of "the biology of blood transfusions."[38] In late May, Bruskin delivered a report to the Eighteenth Congress of Russian Surgeons, detailing the results of twenty-six direct blood transfusions from human to human he had performed to date. Bruskin's efforts culminated in a 170-page monograph on the subject issued by the Narkomzdrav press and an apparatus of his own design for direct blood transfusion, which a Narkomzdrav factory began manufacturing a few months later (Bruskin 1927).

As the aforementioned attempt by Nikolai Kukoverov to establish a blood transfusion institute in Odessa indicates, the operation had quite a few proponents in the Ukraine. Indeed, as early as May 1922, Evgenii Kramarenko, an assistant professor at the surgical clinic of the Odessa Medical School, performed a successful blood transfusion to a four-year-old child suffering from a bad case of scurvy. Independently from Shamov, Kramarenko developed his own methods of producing standard sera for blood typing and organized an exchange of sera with Shamov's group to compare its effectiveness. Reportedly, the sera produced in Petrograd and Odessa proved identical. During the subsequent years, together with his graduate student, Leon Barinshtein, Kramarenko also developed original equipment—needles and canulae—for the operation and regularly employed it in his practice (Kramarenko and Barinshtein 1924; Barinshtein 1928). The first 1924 issue of the influential journal *Modern Medicine* carried a long survey on "blood transfusion in light of current data" prepared by Barinshtein (1924). In April 1924, *Izvestiia* reported a particularly remarkable instance of "miraculous" recovery after a blood transfusion performed by the Odessa surgeons ("Perelivanie krovi" 1924). In September, Kramarenko presented results of his work on blood transfusions to the regional conference of surgeons.[39] In the spring of 1926, the Odessa enthusiasts began to offer a two-week remedial course for doctors on the "theory and practice of blood transfusions" at the Odessa Medical School, detailing methods of producing standard sera, principles of blood typing, and necessary armamentarium.[40]

Shamov's move to Kharkov—at that time, the capital of the Ukrainian Soviet Socialist Republic—stimulated further diffusion of the new operation among physicians in the Ukraine. During 1924–25, Shamov's colleague at the Kharkov University Medical School, Konstantin Gess-de-Kal've, a well-known surgeon-oncologist, used blood transfusions to treat carbon monoxide poisoning and, together with his assistants, studied changes in blood composition after the transfusions (Gess-de-Kal've and Tutaev 1924; Burlachenko and Gess-de-Kal've 1925). Gess-de-Kal've (1925) also published a booklet with detailed descriptions of blood groups, various transfusion techniques, and necessary equipment, focusing in particular on the use of blood transfusions in the pre- and postoperative treatment of malignancies.[41]

After his move to Kharkov, though busy with running his large clinic, Shamov himself continued research on blood groups and blood transfusions. In 1926, together with Vladimir Rubashkin, director of the Ukrainian Institute of Protozoology, Shamov organized a "Permanent Commission to Study Blood Groups." Supported by the Ukrainian Narkomzdrav, the commission became a clearinghouse for a large research program of "blood mapping" in various ethnic populations and geographical regions throughout the Soviet Union.[42] Under the joint editorship of Rubashkin and Shamov, the commission established the first periodical devoted solely to the issues of blood groups and blood transfusions—the *Bulletin of the Permanent Commission to Study Blood Groups*—published simultaneously in Ukrainian and German.[43]

These developments demonstrate convincingly that during the early 1920s, a number of individual practitioners and several research groups were actively engaged in blood transfusions in Russia. Inspired by the successes of their Western colleagues, they elaborated original techniques, perfected methods of blood typing, created the necessary equipment, manufactured standard sera, and publicized the uses of blood transfusions in surgical and therapeutic practice. They reported their findings to professional meetings and published them in professional periodicals. From 1921 onward, medical journals regularly carried literature surveys and original research publications on various issues related to blood transfusions. Beginning in 1923, the popular press joined in by publishing a number of articles on the "lifesaving" technique.[44] By the time Narkomzdrav issued its decree establishing the Institute of Blood Transfusion in Moscow in the spring of 1926, a sizable community of transfusiologists had emerged in Soviet Russia, with a number of eminent surgeons and therapists advocating the adoption of the procedure in medical practice and lobbying for the creation of specialized research institutions. These developments clearly reflected Russian scientists' continuous contacts with their Western colleagues, as well as their untiring efforts to use the newborn Soviet state to advance their own intellectual and institutional agendas, particularly in creating "big science" research institutions.

So how could it happen that Alexander Bogdanov, who in the preceding years had not published a single scholarly article on the subject of blood transfusions, was appointed to head the new institute? As we have seen, in the early 1920s, the Narkomzdrav Scientific Medical Council had opposed the creation of a special institute of blood transfusion. The majority of Soviet transfusiologists worked in Leningrad and the Ukraine, far removed from Narkomzdrav's central offices in Moscow, which considerably impeded their ability to convince the skeptical experts in the Scientific Medical Council and/or Commissar Semashko personally of the necessity of organizing a special institute for the study of blood transfusions. Furthermore, most of them were practicing surgeons interested first and foremost in the practical applications of blood transfusions rather than experimental research. The main proponent of the operation in Moscow, Iakov Bruskin, although also a practicing surgeon, was interested in pursuing experimental research. But he was a junior scientist, and despite his overtures toward the new Bolshevik rulers (in 1924 Bruskin even joined the "Leninism in Medicine" society and published several articles arguing for a "Marxist" approach to medicine), he did not have enough clout to lobby successfully for the creation of a special institute for his own research. The establishment of such an institute obviously required a powerful patron able to overrule the Scientific Medical Council's opposition and a skillful client able to find and court such a patron. Alexander Bogdanov and Joseph Stalin formed a successful, if somewhat reluctant, duo.

3 Voyagers

TO MARS AND BACK

From the tender age of thirteen onward, Alexander Bogdanov was in constant motion throughout his life. He lived in different cities and countries, traversed different professions and fields of knowledge, and steered through safe houses, editorial offices, lecture halls, and operating rooms. In his imagination, he even traveled between planets, between the past and the future, as well as between different economic systems. What propelled this man on his various forays into the unknown in so many different domains? What maps did he read to navigate through unfamiliar territories, and what dictionaries did he rely on to communicate with the locals? What beacons called to him through the fog and darkness of strange lands? If asked, Bogdanov himself would probably have answered these questions with one word: *"nauka,"* the Russian word for science. In a way, he did so in his poem "A Martian Stranded on Earth," written in October 1920 and published as a supplement to a 1924 edition of his science-fiction novel *Red Star.* In this "autobiographical" poem, a lonely Martian is stranded on Earth, all his comrades having died in the crash of their vessel. He cannot return to his planet, and it is "the voice of dispassionate science" that saves him from despair and thoughts of suicide (Bogdanov 1924b, 168). Science was indeed a major guiding light of Alexander Bogdanov's life.

THE REVOLUTIONARY

Alexander was born on August 10, 1873, into the family of a schoolteacher, Alexander Malinovskii, in a small provincial town in western Russia.[1] Very early on, the boy demonstrated a remarkable aptitude for learning. Despite the family's strained financial situation (Alexander had five siblings), at the age of thirteen, he was sent—on a state stipend—to a classical gymnasium in Tula, a large industrial center south of Moscow. Upon graduation from the gymnasium with the highest honors—a gold medal—in the fall of 1892, he enrolled in the Moscow University School of Natural Sciences. Alexander was interested in mathematics, physics, and biology, but he was unable to complete his studies.

In December 1894, along with a large group of other students, Alexander was arrested for "revolutionary activity," expelled from the university, and exiled to Tula, where he began a career as a professional revolutionary (White 1981). He joined a circle of like-minded young *intelligenty,* inspired by the populist movement to further the "enlightenment of the people."[2] Since Tula was an industrial center, renowned for its munitions, bread-making, and metal industries, Bogdanov's educational efforts focused not on peasants, who were the main target of the populist enlightenment campaign, but on factory workers. He began teaching at workers' schools and learning circles. This work inspired Bogdanov's move from populism to Marxism, which considered the proletariat the leading force in the future socialist revolution. In 1897, he published the lectures he had been delivering at workers' schools as a *Short Course of Economic Science,* which became the most popular textbook on Marxist political economy in the early twentieth century, with nine printings appearing in as many years (Bogdanov 1897).

Even though he himself was already teaching, Alexander was determined to continue his own education. Since the conditions of his exile forbade him from residing in the Russian capitals, St. Petersburg and Moscow, as well as the majority of the country's university centers, in the fall of 1895, he enrolled in Kharkov University. But this time, perhaps influenced by the example of his older brother, Alexander chose the medical school. In Russian universities, attending lectures was not obligatory, and he stayed only four to five months in Kharkov to take exams and whatever practical courses were required, spending the rest of his time in Tula deeply engaged in teaching and organizing among workers of the city's famous munitions factory. He was obviously able to get credit for at least some of the general courses he had taken at Moscow University, since he graduated in just four years, in 1899, receiving the diploma of physician with specialization in nervous and psychiatric diseases. After graduation, he married a fellow revolutionary, Nataliia Bogdanovna Korsak (1865–1945), whose patronymic he had adopted as his pen name—Bogdanov.

Also in 1899, Bogdanov published a 250-page-long treatise entitled *Basic Elements of a Historical Outlook on Nature: Nature, Life, Psyche, Society.* Students of Bogdanov's life and works have regularly mentioned this book as the first manifestation of Bogdanov's interest in Marxist philosophy and as a foundation of his own philosophical doctrine, which he would later elaborate in his "empiriomonism," "philosophy of living experience," and "general organizational science."[3] But they have largely ignored other sources of—and, most interestingly, possible reasons for—his first attempt at philosophical theorizing.

In fact, the book presented a peculiar mix of Herbert Spencer's positivism, Ernst Haeckel's monism, Charles Darwin's evolutionism, and Karl Marx's historical materialism.[4] Indeed, it seems that Bogdanov tried to develop a synthesis of these four concepts, somewhat similar to Herbert Spencer's "synthetic philosophy." During the last decades of the nineteenth century, Spencer, Darwin, and Haeckel enjoyed enormous popularity in Russia. Practically all of Spencer's works were translated and published in Russia in numerous editions.[5] Indeed, as Shoshana Knopp (1985) has convincingly demonstrated, Spencer attained something of a "cult status" in 1890s Russia. Spencerian positivism left a deep impression on Bogdanov's thinking, as it did on many other Russian intellectuals of the time.[6] During the same period, Darwin's theory of evolution fueled a lively debate that spread far beyond the biological community, drawing in philosophers, theologians, political commentators and ideologues, sociologists, physicians, writers, and economists. The polemics between proponents and critics of "Darwinism" unfolded in academic journals, novels, and popular magazines, and raged at scholarly conferences, proceedings of learned societies, and dinner parties in private homes. By the 1890s, all of Darwin's works had been published in Russian and the majority of Russian naturalists (though not philosophers) came to consider themselves "Darwinists." Biological evolution became an established fact, even though debates over its concrete mechanisms continued in Russia as they did elsewhere (Vucinich 1988; Todes 1989). Similarly, Ernst Haeckel's works appeared in numerous Russian translations and editions.[7] His "monistic" explorations of life phenomena—from *General Morphology* to the *Natural History of Creation*—generated a lively discussion in Russian professional and popular publications. Haeckel rejected the dualism of "spirit" and "matter," advocating instead "monistic" materialism as the only epistemological principle acceptable to a student of nature.[8] By the 1890s, many Russian naturalists had adopted Haeckelian monism as a basic philosophy of their investigations. As Mark B. Adams (1989, 11–12) has convincingly argued, Bogdanov was certainly familiar with Haeckel's major works.

Thus, in 1899, Bogdanov's forays into Spencerian positivism, Darwinian evolutionism, and Haeckelian monism were far from unique. What made them original was the inclusion of Marx's historical materialism into a "universal," monistic, positivist explanation of the evolution of nature. Bogdanov attempted to demonstrate that "historical outlook" holds the key to understanding all kinds of development processes in nature—from inanimate matter to human society. Darwin's main analytical categories of "selection" and "adaptation" provided a foundation for Bogdanov's analysis. He suggested that selection and adaptation operate in the same way in nature, thought processes, and society. Bogdanov's *Basic Elements* first articulated the idea that certain "general principles" could be discerned in nature at all levels of its organization, including inanimate and animate matter, individual minds, ideological constructs, and social institutions—an idea he would elaborate throughout his life.

But why would a newly minted psychiatrist and an active Marxist propagandist, who had already made a name for himself in political economy, write a philosophical tract? Bogdanov had obviously written the book during his last year of studies in Kharkov, and it seems likely that he had planned to submit this work as a dissertation for a degree in philosophy or psychology (which was at that time taught as a part of philosophy).[9] Certain features of Bogdanov's text support this hypothesis: it was definitely structured according to the standard form of a dissertation. Opening with a general introduction that provided definitions of its main analytical categories, the text consisted of four main chapters and a conclusion. It also included a thirty-page "appendix" that discussed the basic "elements" of psychology, such as sensations, emotions, mind, consciousness, unconsciousness, perceptions, and memory.

Furthermore, two of Bogdanov's university professors seem likely candidates for advising, inspiring, and encouraging him in this endeavor. First was his psychiatry teacher Iakov Anfimov (1852–1930).[10] Anfimov received his first degree from the St. Petersburg University School of Natural Sciences and a second from the Military-Medical Academy. For several years, he worked as a physician and at the same time prepared a doctoral dissertation on the anatomy and physiology of the central nervous system. In 1887, after the successful defense of his dissertation, he became a lecturer (*privat-dozent*) at the VMA department of nervous and psychiatric diseases. In 1892, he left St. Petersburg to take up a professorial position at Tomsk University and, two years later, moved to Kharkov to chair the department of nervous and psychiatric diseases. Anfimov's professional interests were quite diverse and encompassed not only psychiatry but also anatomy, physiology, psychology, sociology, and philosophy. Reading voraciously, he was well informed of the newest developments in all these fields and demanded the same from his students.[11] He approached psychiatric problems from broad social, philosophical, biological, and psychological perspectives and developed a synthetic understanding of the human psyche, clearly articulated in his 1893 book *Mind and Personality in Psychiatric Disorders* (Anfimov 1893). Obviously, in his own work, Bogdanov tried to expand further the synthesis attempted by his teacher.

Bogdanov's second, and perhaps even more influential, teacher was Fedor Zelenogorskii (1839–1909).[12] A graduate of the renowned Kazan University, Zelenogorskii made his career in Kharkov, eventually becoming the head of the philosophy department. He wrote prolifically on various subjects in the history of philosophy, from Plato and Aristotle to Immanuel Kant and Friedrich Schelling. He was also quite interested in psychology, and Bogdanov was undoubtedly familiar with his major work in this field, the *Outline of the Development of Psychology from Descartes to the Present Day* (Zelenogorskii 1885). But it was Zelenogorskii's doctoral dissertation "On Mathematical, Metaphysical, Inductive, and Critical Methods of Investigation and Proof," which pointedly discussed the origins and developments of, as well as differences and similarities between, various methods of reasoning and explanation, that appears to have been a major inspiration for Bogdanov's *Basic Elements* (as well as for many of his later writings).[13] Bogdanov's work shows that he took to heart his teacher's definition of philosophy's major goal: the "*inference, generalization, and critique*" of new scientific discoveries and methods in order to create a "common method" for all sciences (Zelenogorskii 1998, 82, italics in the original). Indeed, in *Basic Elements,* Bogdanov supplemented Zelenogorskii's dissertation by analyzing one "method of investigation and proof" that had been absent in his teacher's treatise— the historical method—and demonstrated its application to a range of sciences from physics to sociology. In doing so, Bogdanov followed in the footsteps of yet another professor, whose works he had encountered while studying at Moscow University, Kliment Timiriazev (1843–1920).[14]

A prominent plant physiologist who had won international acclaim for his studies on the mechanism of photosynthesis, Timiriazev was one of the most popular professors at Moscow University, and his lectures always attracted large crowds of students. He was particularly famous as Russia's "Darwin's bulldog" for his untiring defense of Darwin's evolutionary theory. In the winter of 1890–91, Timiriazev delivered a series of public lectures on "Darwinism," which were a fixture of Moscow's cultural life that season and had great resonance among the educated public. Although Bogdanov could not possibly have attended the lectures, in the fall of 1892—exactly at the time when Bogdanov entered Moscow University—the influential monthly *Russian Thought* began publishing Timiriazev's lectures under the general title "Historical Method in

Biology" (Timiriazev 1892–95). Timiriazev argued that only with Darwin's introduction of "historical method" did biological investigations—from morphology and systematics to physiology and embryology—become truly scientific, moving from simple observations to causal explanations of biological phenomena. He advanced the same argument in his four-hundred-page volume *Charles Darwin and His Theory*, published in the summer of 1894, which became a definitive Russian text on Darwinism for years to come (Timiriazev 1894a). Furthermore, as witnessed by his critique of vitalism published the same year in *Russian Thought*, Timiriazev (1894b) also was an avowed "monist."

Although he did not directly refer to Timiriazev in *Basic Elements*, Bogdanov's later works demonstrate unambiguously that Timiriazev influenced profoundly not only his biological views and philosophical interests but also his attitudes toward science more generally. In his eulogy at Timiriazev's grave in April 1920, Bogdanov praised him as an "untiring fighter for the scientific worldview," "who had done more than anybody else for the victory of Darwinism and scientific materialism" (Bogdanov 1920d, 1–3).[15]

It is unclear from available materials why Bogdanov did not submit *Basic Elements* as his dissertation. Perhaps Zelenogorskii's retirement from Kharkov University in the spring of 1899 (exactly at the time when Bogdanov completed his tract) played a role. More likely, Bogdanov's revolutionary work interfered with his philosophical pursuits. In November 1899, Bogdanov was again arrested for spreading revolutionary propaganda among workers. After several months in prison, he was exiled to Kaluga, a provincial town not far from Tula. A few months later, he was sent much farther north to the city of Vologda to serve his four-year sentence. There, he joined a group of exiles that included, among others, the future Bolshevik commissar of enlightenment Anatolii Lunacharskii (who would later marry one of Bogdanov's sisters) and philosopher Nikolai Berdyaev, a prominent critic of Bolshevism.[16] In Vologda, Bogdanov spent his days making rounds in the local psychiatric asylum and his nights writing on—and debating with other exiles—the future of his country.[17] But he did not abandon his philosophical interests. In 1902, in an influential monthly, *Education*, Bogdanov published an abridged but updated version of his "synthesis" of Marxism and Darwinism under the title "The Development of Life in Nature and Society" (Bogdanov 1902). During the next two years, in the leading Russian philosophical journal of the time, *Questions of Philosophy and Psychology*, Bogdanov published a series of articles expanding and updating his "monistic" views on psychology, interrelations of psychic and biological evolution, and "ideals of cognition" (Bogdanov 1903a, 1903b, 1904b).

In the spring of 1904, when the term of his exile ended, Bogdanov left Russia for Switzerland. There, he joined Lenin in the fierce internal debates within the Russian Socialist-Democratic Labor Party, which had led to its split into two competing factions: the Bolsheviks (literally, "the majority") and the Mensheviks (literally, "the minority"). Together with Lenin, he became a member of the Bolsheviks' highest council—the Bureau of the Majority Committees. Bogdanov also served on the editorial boards of the party's major periodicals and published extensively on the tactics and strategies of the Russian revolutionary movement. In the fall, he returned to Russia and joined the St. Petersburg committee of the Bolshevik Party, working on the preparations for the party's general congress, which was to convene in London the next year. Bogdanov came to London and delivered two key reports to the congress: on the armed uprising and on the party's internal organization. In a clear recognition of his contributions, the congress elected Bogdanov to the party's ruling body, its Central Committee.

At the time of the first Russian revolution, of 1905–6, Bogdanov represented the Bolsheviks in the St. Petersburg Soviet of Workers' Deputies. Along with other members of the Soviet, in December 1905 Bogdanov was arrested, imprisoned, and then exiled from Russia. He returned to Russia illegally, and while living (together with Lenin) in Finland, less than a hundred miles from St. Petersburg, he took an active part in the political struggles of the time, publishing on a variety of issues in the party's periodicals. In late 1907, together with Lenin, Bogdanov again moved abroad to become a lead editor of the party's main journal, *The Proletarian.*

During these busy years, Bogdanov did not limit his work exclusively to the pressing issues of day-to-day practical politics and revolutionary propaganda. He also became deeply engaged in elaborating Marxist theory. Bogdanov took on the task of extending Marx's class analysis from capitalism's economic "foundation" to its "superstructure." He focused in particular on "culture" and paid scant attention to other elements of the superstructure, such as politics, laws, bureaucracy, or social institutions. For Bogdanov, "culture" was the manifestation of class consciousness and included not only the arts but also specific patterns of human behavior and thinking: ideology, philosophy, and science. Following Marx, Bogdanov considered culture in all its forms as an extension of labor processes. Even language, according to Bogdanov, emerged and evolved within and alongside the evolution of labor.

As Marx in his analysis of the capitalist economic system had discerned certain elements of a future socialist economy, so too in his analysis of "bourgeois" culture, Bogdanov glimpsed certain elements of a future socialist "proletarian" culture. According to Bogdanov, the fundamental difference between "bourgeois" and "proletarian" cultures was that the former is entirely defined by *individualism* inherent to a capitalist economy due to the private ownership of the means of production. In contrast, the latter is intrinsically defined by *collectivism,* since proletarians by definition do not own anything but their labor, and they are forced by the conditions of factory production to develop and embrace collectivism. Bogdanov attempted to elaborate theoretically the concept of proletarian culture, writing extensively on the subjects of proletarian ideology, epistemology, ethics, aesthetics, and science.

The questions of proper "proletarian" philosophy, particularly epistemology, occupied a large place in his theorizing. Building upon the works of the prominent Austrian scientist-turned-philosopher Ernst Mach, the German chemist Wilhelm Ostwald, and the German philosopher Richard Avenarius, he formulated an original epistemological concept of "empiriomonism" (Bogdanov 1904-6).[18] He also produced Marxist analyses of human psychology, the "psychology of society," and "the goals and norms of human life" (Bogdanov 1904a, 1904c, 1905a, 1905b).

In 1909, Bogdanov's disagreements with Lenin on a number of tactical and theoretical issues led to his expulsion from the editorial board of the Bolshevik Party's oracle, and a few months later, from its Central Committee.[19] Lenin (1909) even published a four-hundred-page-long treatise entitled *Materialism and Empiriocriticism,* specifically to refute Bogdanov's "reactionary" philosophical ideas and to undermine his reputation as the party's leading theoretician. Bogdanov (1910) responded in kind, publishing a year later under the title *Faith and Science* his own denigrating analysis of Lenin's critique. For a time, Bogdanov attempted to lead an independent faction, and, with the financial support of the eminent Russian writer Maxim Gorky, he organized party schools for young Russian revolutionaries in Italy.[20] Soon thereafter, however, Bogdanov seemed to have lost interest in practical politics and focused primarily on his writings.[21]

THE WRITER

It was during the time of his most intense work on "proletarian" culture and his growing disagreements with Lenin that Bogdanov first tried his hand at fiction. In 1908, he published (in Russia and at his own expense) a novel, entitled *Red Star: A Utopia,* which depicted the Martian adventures of a young professional revolutionary from Earth, appropriately named Leonid.[22] The novel presented in fictional format

3.1 Lenin and Bogdanov in Italy, circa 1908. *Sitting left to right:* Vladimir Bazarov (Rudnev), Vladimir Lenin, and Alexander Bogdanov; *standing:* Maxim Gorky, Gorky's son Maxim Peshkov, and Nataliia Korsak.

the ideas about socialism Bogdanov was at that very time elaborating in his theoretical writings.[23] Apparently, a main reason Bogdanov wrote the novel was to make these ideas more accessible to his intended audience—"conscientious workers." But one cannot exclude the possibility that Bogdanov also wanted to explore these ideas without the strictures of a formal theoretical treatise, allowing his imagination to fly over gaping holes in his analysis of what a socialist society ought to be. It was on Mars that Bogdanov envisioned a future socialist society and a future for blood transfusions.

As many scholars have noted, Bogdanov used the stock devices of a classic utopia and combined them with the standard trappings of science fiction (SF), a genre that was at that very time capturing the attention of readers in many countries, including Russia.[24] As in a classic utopia, Bogdanov's novel described—through the eyes of a "naïve" traveler—an ideal society located in a faraway land and populated by "advanced" human beings.[25] Like so many other SF writings of the time, it was set on Mars and contained lengthy didactic depictions of Martian scientific, technological, social, and biological advances.[26]

The originality of *Red Star* that distinguished the novel from its literary predecessors and contemporaries derived from its author's Marxist convictions and scientific/medical training.[27] Unlike classical utopias, Bogdanov's presented not the ultimate "golden age"—a static, free of conflict and strife "paradise"—but a dynamic society that continues to develop by resolving its conflicts and contradictions. Furthermore, the novel's subtitle—probably for lack of a better word to define its genre at the time—did not do justice to its substance and more important to Bogdanov's actual vision that it conveyed.[28] As Bogdanov himself would explain a few years later in a "prequel" to *Red Star*—which quite tellingly was subtitled not "utopia" but "fantastic novel"—"utopia expresses strivings that cannot be realized, efforts that are insufficient to overcome obstacles" (Bogdanov 1913b, 99). But while writing *Red Star*, Bogdanov himself had no doubt that the strivings expressed in his "utopia"—a socialist society—could be, indeed inevitably would be, realized and that, in his view, humanity already possessed the instrument for overcoming any obstacles: *science*. Unlike many SF writings of the day, Bogdanov's novel used "wondrous" science and technology not merely as ends in themselves or as simple devices of "estrangement," but as a central component of his vision, exploring the role science could and should play in the advanced socialist society he was portraying.

According to the ideas of what a socialist society ought to be that Bogdanov was developing in his theoretical writings, his Martian society is a direct reflection of its economic foundation. As the result of a socialist revolution that occurred some two hundred years earlier, there are no states, private property, family, money, law, and politics in any earthly sense on Mars. All Martians speak one common language. Labor is completely voluntary, and economy is managed scientifically by a "central statistical bureau" equipped with something akin to computers to provide to all and each according to their needs.

The defining feature of Martian society is *collectivism:* every facet of its life—its arts, clothing, education, interpersonal (including sexual) relations, medicine, consumption, science, decision making, technology, everyday life—is completely devoid of the individualism to which the visitor from Earth is so accustomed. Even Martians' physical appearance bears almost no signs of individuality, and it takes Leonid a while to learn to distinguish one Martian from another. There are no obvious distinctions between the sexes (even Martian language seems to completely lack gender qualifications) and no signs of aging. Indeed, we learn that Leonid experiences a "strange attraction" to Netti (a Martian doctor he has befriended during his trip to Mars) but feels thoroughly confused about it, because he does not realize that Netti is a woman, until—near the end of the story—she tells him directly, and the love affair ensues. Furthermore, as Leonid discovers, collectivism has imbued not just Martians' psychology and behavior, but even their physiology.

Among his various exploits on Mars, Leonid studies science, arts, and literature; works at a factory; visits schools and hospitals. But in every sense he feels inferior to his hosts. He has great difficulties studying Martian science, for its very logic seems incomprehensible to him. He is unable to truly enjoy Martian literature, theater, paintings, sculpture, or architecture (strangely enough, there is no mention of music or dance), even though he seems to understand, on a rational level, their beauty. He cannot keep up with the tempo of work in the factory, for his attention span, power of concentration, and reaction time are all significantly lower than those of an average Martian. Frustrated, suffering from insomnia, and feeling on the verge of a nervous breakdown, Leonid decides to see a doctor, Netti.

In the best tradition of SF, his visit to the doctor occasions a long exposition of Martian views on life, death, and medicine (Bogdanov 1908, 79–85). Among many other questions, Leonid asks Netti: "Why do you Martians preserve your youth for so long? Is this a peculiarity of your race or the result of better living conditions, or is it something else?" In a typical for SF "lecture" format, Netti patiently explains to the Earthling that "race has nothing to do with it: two hundred years ago our life span was one half of what it is now. Better living conditions? Yes, to a certain degree, but only to a degree. The main role is played by the renewal of life that we use."

The "renewal of life" turns out to be the exchange of blood among Mars's inhabitants. "You know that in order to increase viability of cells or organisms, Nature constantly supplements one individual with another," Netti continues, "For this purpose, two unicellular organisms conjugate into one, and only in such a way, they regain the ability to procreate—the 'immortality' of their protoplasm. The same principle is behind the sexual crossing of higher plants and animals: here also the vital elements of two different beings unite in order to create a more perfect embryo of the third one. Finally, you also already know the use of blood sera to transfer the elements of viability from one being to another, so to say, piecemeal—for instance, as the increased resistance to a particular disease. We've taken this further and conduct blood exchanges between two human beings, each one of whom can give to another a mass of elements that increase viability.

"It is simply simultaneous blood transfusion from one individual to another and back accomplished by connecting their blood vessels through special apparatuses," Netti elaborates. "If all precautions are taken, it is perfectly safe; the blood of one person continues to live in the organism of the other, mixing with his own blood and profoundly renewing all his tissues."

Leonid is puzzled: "Are you able to rejuvenate old people in such a way—by transfusing young blood into their veins?"

"To an extent, yes," answers Netti, "but, of course, not altogether, for blood is not everything there is in the organism, and [transfused] blood, in its turn, is reworked by the body. That is why, for example, a young person does not age from the blood of an old one: whatever weak and old [elements may be] present in the blood, [they] are quickly overcome by the young organism, which at the same time absorbs from the [old] blood many elements that it lacks; the energy and flexibility of its vital functions also increase."

Leonid is surprised: "But if this is so simple, why does not our earthly medicine use this method yet? After all, blood transfusions have been known in our medicine for several hundred years, if I am not mistaken."

Netti claims ignorance: "I don't know. Perhaps there are some specific organic conditions that on Earth render the method ineffective." But she ventures an explanation: "Perhaps it is merely a result of your dominant psychology of individualism, which so deeply isolates people from one another that the thought of their vital fusion is almost incomprehensible to your scientists. Besides, there are among you many diseases that poison blood—diseases often unknown to, or concealed by, those who have them. The blood transfusions practiced—though now very rarely—by your medicine have a somewhat philanthropic character: a person with much blood gives some of it to another person who desperately needs it due to, say, profuse bleeding from a wound. It happens here, of course, but what is constantly in use is something else—something that corresponds to the nature of our entire system:

the comradely exchange of life not only in the ideal but also in physiological existence."

The issue of collectivism was certainly central to Bogdanov's understanding of socialism and its future proletarian culture. Indeed, a year after he had published *Red Star*, Bogdanov compiled a special volume, entitled *Outlines of the Philosophy of Collectivism*, with contributions from his closest friends and followers who shared his ideas: Vladimir Bazarov, Anatolii Lunacharskii, and Maxim Gorky (*Ocherki . . .* 1909). Bogdanov (1909a, 1909b) himself wrote two pieces for the volume: one on the interrelations of "science and philosophy" and another—more than a hundred pages long—on the "philosophy of the modern scientist." Tellingly, for his first article in this collection, Bogdanov used the pen name "Verner," the same pen name he had used in the *Red Star* for Leonid's friend and publisher of his manuscript. But how did Bogdanov move from philosophy to physiology and to biology more generally? And how did he come up with these strange ideas regarding blood exchanges?

The "hospital episode" in *Red Star* demonstrates that Bogdanov read attentively popular accounts of various subjects in contemporary experimental medicine and biology. He surely knew of Emil von Behring's work on "serum therapy" for diphtheria, which was awarded the first Nobel Prize in medicine in 1901. He was definitely familiar with the theories of immunity developed by his compatriot, zoologist Elie Metchnikoff, and the German bacteriologist Paul Ehrlich, who shared that year's (1908) Nobel Prize. But in all likelihood Bogdanov did not follow the specialized literature of the day on blood chemistry and blood transfusions. The "hospital episode" shows no knowledge of the existence of different blood groups discovered by Karl Landsteiner in 1901 (or Jan Jansky's 1906–7 work that refined Landsteiner's findings).[29] Nor does it convey any familiarity with the sophisticated surgical techniques for direct blood transfusions that George W. Crile (1907) and Alexis Carrel (1907) had developed by that time.

Many ideas expressed in *Red Star* regarding biological differences between Martians and Earthlings reflect Bogdanov's general understanding of contemporary biology, particularly on issues of heredity, variability, and evolution. As we have seen, already in his first philosophical work, Bogdanov had considered *selection* a "basic element" of all kinds of historical development. Together with his main biology teachers—Haeckel and Timiriazev—Bogdanov certainly subscribed to the view that natural selection was the principal causal factor of biological evolution. Yet, as did many of his compatriots, he rejected the notion of intraspecies competition—embodied in the struggle for existence—as the mechanism underpinning natural selection. For him, as for many other Russian biologists, the struggle for existence meant the struggle with unfavorable environmental conditions, not the Malthusian/ Darwinian competition among members of the same species.[30] Indeed, in *Red Star*, the struggle against "nature"—building canals to expand agricultural production and searching for necessary resources on other planets—constitutes the leading factor of continuing social evolution on Mars. Martians refuse to practice birth control (Bogdanov 1908, 77), which many of Bogdanov's contemporaries regarded as the only means to decrease the competition for resources on Earth. And it is through cooperation, not competition among the Mars inhabitants, as in Edgar Rice Burroughs's novels, that they are able to "conquer" the nature of their planet and begin the conquest of other planets.[31]

Bogdanov undoubtedly read the work of his compatriot Prince Petr Kropotkin—the leading theoretician of anarchism living at the time in political exile in England—on mutual aid as a factor of biological and social evolution.[32] Indeed, there are many similarities between Kropotkin's descriptions of mutual aid among human beings and Bogdanov's depictions of collectivism in both his fictional and nonfictional writings. Yet as a convinced Marxist, Bogdanov could not and did not accept Kropotkin's notion of mutual aid as a *universal* factor of biological and social evolution.[33] For him, the capitalist economic system "automatically"—this was one of Bogdanov's favorite examples of the influence of economic foundation on superstructure—produced fierce competition among individuals and thus promoted the individualism so characteristic of bourgeois culture. Only with the victory of socialism could the elements of collectivism, which began to develop among the proletariat under capitalism, emerge triumphant! And this is exactly what we see in *Red Star*.

Already his *Basic Elements,* and especially his 1902 "The Development of Life in Nature and Society," demonstrated Bogdanov's deep interest in the interrelations among evolutionary processes unfolding on biological, psychological, and social levels. In *Red Star*, he clearly attempted to explore these relations further. On Mars, the socialist revolution precedes and "instigates" biological and psychological evolution, leading—in the course of just two hundred years—to the emergence of a number of new biological and psychological characteristics among its inhabitants. Compared to humans in terms of their psychology, Martians demonstrate much faster reaction time, much better control over their emotions, and much more developed cognitive abilities. Martians' new biological features include enlarged eyes and skull (implying larger brains), lower susceptibility to disease (there is no mention of infectious diseases at all), a doubled life span, greatly diminished differences between males and females, and a dramatic decrease in individual variability in outward appearance. And apparently all these new traits appeared in a very short time: as Leonid notices during his visit to the Martian Museum of Fine Arts, in the statues of the "capitalist period, sexual differences are expressed much more strongly" (Bogdanov 1908, 70).[34] The possible mechanisms of these rapid biological and psychological changes, however, remain unclear. Certainly, Netti's lecture implies that "comradely exchange of life" played a decisive role in bringing these changes about—at least as regards the Martians' extended life span and lowered susceptibility to disease. But Netti/Bogdanov did not explain exactly how blood transfusions could have produced this change or whether the new biological and psychological traits were hereditary.

In *Basic Elements*, Bogdanov had clearly sided with the concept of "blended heredity" and the Lamarckian notion of the inheritance of acquired characteristics then prevalent among biologists and physicians (Bogdanov 1899, 84–88).[35] But since he had written that book, the science of heredity had advanced dramatically. In 1900, Gregor Mendel's concept of "nonblended" heredity had been rediscovered, and during the early 1900s, a number of biologists actively popularized "Mendelism" as a foundation for the new science of genetics (Olby 1966). Although nowhere in the novel did Bogdanov state explicitly his views on heredity, nothing in his writing indicates that he was familiar with Mendel's concept. But a few statements scattered in *Red Star* suggest that along with many of his contemporaries, including both of his biology teachers, Haeckel and particularly Timiriazev, who was quite critical of Mendelism,[36] Bogdanov still believed in the inheritance of acquired characteristics. The "hospital episode" shows that Bogdanov knew of, but largely misunderstood, the highly publicized experiments with infusoria by the continental cytologists Otto Bütschli and Emile Maupas.[37] Both Bütschli and Maupas tried to prove experimentally the idea of the "immortality of germ plasm" advanced by the influential German biologist August Weismann as an alternative to the Lamarckian inheritance of acquired characteristics.[38] But, as did many of his contemporaries, Bogdanov misread these experiments as a proof of the "immortality of their [infusoria's] protoplasm," which meant that infusoria as living organisms were virtually "immortal" but said nothing on the question of heredity. Bogdanov seemed to imply that blood exchanges among Martians actually "leveled out" their individual and sexual differences and extended their life span (not to mention their

psychological abilities) through the inheritance of acquired characteristics. His later writings did raise such a possibility explicitly.

Although the "hospital episode" was just that—an episode in a much
larger work about a future socialist society, the idea of blood exchanges
it had sketched out obviously continued to intrigue its author. Sometime
in 1910, two years after the publication of the novel, Bogdanov penned
(in French) a seven-page program of investigating "the comradely exchange of life."[39] The program elaborated theoretical arguments and
practical methodology for what Bogdanov now termed "La greffe du
sang: Méthode l'échange." It shows that Bogdanov had familiarized himself with contemporary research in tissue and organ transplants (at the
time, often referred to as grafts), as well as with available instruments
and methods of blood transfusion. He even thought of a few improvements to the existing apparatuses for direct transfusions marketed by
British and French manufacturers.

The essence of Bogdanov's exchange method was "direct, simultaneous, and mutual transfusion of blood between two individuals of
the same species." Bogdanov suggested that two apparatuses could be
used simultaneously to link up the individuals' "corresponding veins,"
thus increasing substantially the volume of transfused blood. He even
thought that "one could connect successively the same subject with
several others; thus almost the whole mass of his blood will be renewed." Obviously, the existing methods of serotherapy, which Bogdanov characterized as nothing more than "a transfusion with dead and
modified blood," provided the main inspiration for his ideas. Caution-
ing about the possible risks involved, he envisioned several types of
experiments in animals and humans, which, in his view, would solve
important problems facing contemporary biology and medicine, particularly the transmission of immunity through "live blood" and the
nature of "physiological individuality, which is apparently much more
expressed in humans than in animals." Exclaiming rhetorically "what
other medium than blood and lymph would transmit to the generative
cells the stimuli of corresponding variation," Bogdanov thought that
his method would clarify the question of the inheritance of acquired
characteristics, which boggled the minds of many biologists and physicians at the time. He was obviously unaware that nearly forty years
earlier, Francis Galton (1870–71) had conducted extensive experiments
demonstrating that blood transfusions did not transfer hereditary traits
from one rabbit to another. Following the ideas presented in *Red Star,*
Bogdanov also thought that his method "would give an experimental
resolution to alchemists' dreams about the possibility of reinforcing
and rejuvenating the organism by renewing its blood." Bogdanov was

3.2 A page from Bogdanov's 1910 program on "transplantation of blood." (Courtesy of RGASPI.)

sure that experiments with blood exchanges would produce "interesting results, whatever their practical import." But at the time, his "research program" remained buried among his numerous papers.

In 1913, Bogdanov published his second fictional work, but not "a utopia." The "fantastical novel," entitled *Engineer Menni,* was a sort of prequel to *Red Star.* Supposedly written by a Martian and translated for the Earthlings by Leonid, the hero of the first novel, the book depicted the key episode of Martian "presocialist" history—the "Plan of the Great Works"—the gigantic project of building the Martian canals, some two hundred fifty years prior to the events described in *Red Star*. Bogdanov recounts Martian history through the life of three generations of the same family: the "feudal lord," Duke Ormen Aldo; his son, "bourgeois" engineer Menni; and his grandson, a "socialist" engineer named Netti. This simple technique allows Bogdanov to present his Marxist views of history—as a progression from feudalism to capitalism to socialism—and to use the three protagonists to illustrate the profound differences in their "class consciousness." For the scheme to work, Bogdanov makes each Menni and Netti an "orphan," who has not been raised by his "alien-class" father but has imbibed his particular class consciousness from foster parents—"republicans" in the case of Menni and "proletarians" in the case of Netti—thus confirming and conforming to Marx's famous thesis that one's material conditions of life determine one's consciousness. Focusing specifically on the interactions between Menni and Netti, who meet for the first time when Netti is a grown man, Bogdanov demonstrates the utter incompatibility of their worldviews, even entitling one of the chapters: "The Two Logics" (Bogdanov 1913b, 81–89).

But the novel's major theme is not just any form of class consciousness but, first and foremost, "science," *nauka*. The novel is in fact a fictionalized reiteration of Bogdanov's Marxist analysis of "bourgeois" culture applied specifically to science, which explains its often "wooden" prose, cartoonish characters, and weak plot line. Bogdanov presents his views through Netti's diatribes against capitalism at workers' meetings and through lengthy dialogues between Menni and Netti about the Great Works, which make the novel a much less engaging and interesting read than *Red Star*.

In accordance with Russian use, Bogdanov's *nauka* includes not only all natural sciences but also philosophy, social sciences, and the humanities. What is more, in the novel it is synonymous with technology, engineering, and management. Throughout the novel, Menni is characterized interchangeably as a "physicist," a "great scientist," an "engineer," an "inventor," and an "administrator." Similarly, his son is portrayed as a "scientist," a "chemist," an "engineer," an "administrator," and a disciple of "the great scientist Xarma" (an obvious anagram of Marx's name), who elaborates further Xarma's materialist theories of political economy and history.

According to Netti's "Xarmian" analysis, science is an outgrowth, an extension of Labor. But under the conditions of a capitalist economic system, its true function—as the means of organizing labor and production—is perverted. In its relentless pursuit of profits, capitalism employs science for the organization of production, but it separates science from labor. As a result, even in advancing projects that supposedly should benefit everyone (such as building Martian canals), science serves first and foremost the interests of capitalists and itself becomes an instrument of oppression of the proletariat. Blinded by his "bourgeois" idealist worldview, Menni is unable to understand this materialist analysis. For him, the essence of science is Idea. Of course, some "practical" ideas could help humanity conquer and transform nature, as does his own idea of the Great Works, but science itself exists for the sake of science: the purpose of science is discovery, development, and accumulation of true Ideas. Even after Menni himself is imprisoned as a result of capitalists' conspiracy to remove him from the post of chief administrator (since as an honest man he hinders their attempts to extract the maximum possible profits from the Great Works), Menni is incapable of understanding the unbreakable link between science and labor, and the role science plays in capitalist society.

Yet, according to Netti, capitalism perverts not just the purpose but the very nature of science, leading to its fragmentation into numerous specialties and infusing it with individualism inherent to the capitalist "anarchic" mode of production. Each fragment loses its connection with the whole and develops its own esoteric language, making science incomprehensible to the uninitiated—first of all, the proletariat. Deprived of access to science—reinforced by the capitalist system of education—the proletariat en masse is reduced to reliance on *faith* in following its leaders, instead of each proletarian being able conscientiously, using his personal knowledge of science, to understand reality and hence the necessity of particular actions. The proletariat needs its own science, Netti concludes, one—collectivist in its outlook and practice—that is much simpler, more general, and much more closely connected to labor than "bourgeois" science. Indeed, the novel ends with Netti—aided by a group of faithful disciples—working on creating such a universal "proletarian" science and starting its first major project, the "Workers' Encyclopedia," slated to play an important role in preparing the future Martian socialist revolution.

THE SCIENTIST

In his own life, Bogdanov followed the path chosen by his fictional character Netti: he immersed himself in working out a foundation for proletarian science, not on Mars but on Earth—to be exact, in Russia. In 1913, the Russian Empire celebrated the three hundredth anniversary of the Romanov dynasty, and to mark the jubilee, the czar announced amnesty for political exiles. In October of that year, Bogdanov legally returned to Russia and settled in Moscow. He published his ideas about science in future society as a lengthy article on the "secret of science" in an influential monthly, *The Contemporary* (Bogdanov 1913c). He also published a three-hundred-page-long treatise on the "philosophy of living experience," updating and summarizing his views on inter-relations between major philosophical doctrines (idealism, materialism, empiriocriticism, dialectical materialism, empiriomonism) and a "future science" that in his view would eventually replace all of philosophy (Bogdanov 1913a).[40] According to Bogdanov, historically, the principal task of any philosophy was to produce a generalized view of the world. But the only source of true knowledge of the world is "labor practice" and, by extension, science. Since the methods used by philosophy are infinitely inferior to the methods of science itself, when (under social-ism) science overcame its fragmentation and specialization (produced by its capitalist foundation) and focused its investigations on "general principles," science would supplant philosophy in its main task and thus would render any and every philosophy obsolete.

The same year, Bogdanov published the outlines of this "future science": the first volume of his "universal organizational science" or "tectology," as he named it, borrowing the term from Ernst Haeckel (Bogdanov 1913d). Developing several common themes of his previous writings, Bogdanov searched for "general principles of organization" ap-plicable to any and every complex system (whether a planetary system, a human society, a religion, or a beehive) and attempted to analyze the defining features of such a system: unity and wholeness. What makes a set of separate parts into a whole, and how is this accomplished? What keeps the separate parts united, and how can this unity be achieved? What accounts for the whole breaking into its constituent parts? Bog-danov "dissected" various systems to find "generalized" answers to these questions.

Yet immersed in this search for general principles, Bogdanov did not forget his earlier thoughts about blood exchanges. He incorporated the ideas first presented in *Red Star* and further elaborated in the 1910 program into his discussion of the "universal formation mecha-nisms." In a section entitled "Certain Examples of Conjugation Pro-cesses," right after the description of "the conjugation of dialects and languages," Bogdanov inserted a special subsection on "physiological exchange-grafting" (Bogdanov 1913d, 142–48). Emphasizing the role of blood as a unique liquid tissue that unifies all other organs and tissues in the organism, he discussed blood exchanges as a means of increasing the "viability" of participants in such exchanges due to the reciprocal transfer of certain "vital elements"—for instance, immunity to particular diseases. Answering the question posed by his character Leonid in *Red Star,* Bogdanov hypothesized that blood exchanges between the old and the young should indeed lead to "rejuvenation."[41]

Life interrupted Bogdanov's intense theoretical work. With the start of the Great War in the summer of 1914, Bogdanov was drafted into the army as a doctor,[42] since medicine was his "official" profession. For half a year, he served with an infantry regiment on the front lines. But for a man whose entire medical experience was limited to a few months of rounds in a psychiatric asylum more than ten years earlier, his military medical duties proved more than he could bear. In February 1915, Bogdanov suffered a severe nervous breakdown that landed him in the hospital for three months. He recovered but was deemed unfit for frontline duty and was assigned to the staff of a large military hospital in Moscow. Bogdanov served as a "junior intern," but after a year, he was reassigned once more. In the summer of 1916, he became a medical inspector for POW camps located in the Moscow region. This new post was apparently much less demanding in terms of required medical skills and left him enough free time to resume his writings. By the summer of 1917, Bogdanov had finished the second volume of his *Tectology,* elaborating further his ideas of "universal organizational science," and toward the end of the year, he published it at his own expense (Bogdanov 1917a).

After the February 1917 revolution that dethroned Czar Nikolas II, Bogdanov joined the "educational-cultural department" of the Moscow Soviet of Worker Deputies, lecturing and publishing extensively on various issues related to the proletariat's position in the new, democratic Russia (Bogdanov 1917b). He took no part in the October 1917 Bolshevik coup d'état. Furthermore, just a few weeks after the coup, he declined the invitation of his brother-in-law Lunacharskii, who came to head the Commissariat of Enlightenment (Narkompros) in the new Bolshevik government, to accept a post in the new agency in charge of the country's cultural affairs. As his response to the invitation shows, Bogdanov did not want anything to do with his former "brothers-in-arms," the Bolsheviks.[43] Instead, he became very active in the Proletkult (proletarian culture) movement his ideas had helped instigate before the revolution. He was elected to the Proletkult governing council and to the editorial board of its mouthpiece, *Proletarian Culture,* and he immersed himself in putting his theoretical ideas into practice.[44] In February 1918, he delivered a keynote address to the first congress of Proletkult Moscow chapters, entitled "Science and the Working Class." The address propagated Bogdanov's favorite idea that to become true masters of their socialist society, workers needed first to master science, particularly, Bogdanov's own tectology. As the first steps toward reaching this goal, and following the path of his fictional character Netti, he proposed the creation of "proletarian universities" and the publication of a "Workers' Encyclopedia" (Bogdanov 1918e).[45]

A few months later, Bogdanov also became a founding member and a member of the ruling body—the Presidium—of the Socialist (later rechristened "Communist") Academy, established in July 1918 as the center of Marxist learning and scholarship.[46] He began teaching Marxist political economy and philosophy at various educational institutions, including the just-created "proletarian university" and his former alma mater, Moscow University. Also in 1918, Bogdanov secured two new editions of his SF novels, *Red Star* and *Engineer Menni*.[47]

During the years of the civil war, from 1918 to 1921, thanks to his affiliation with the Socialist Academy, Bogdanov was exempted from the military draft as a "very much needed and indispensable specialist."[48] Amid famine, epidemics, and shortages of nearly everything, he continued his theoretical work, publishing extensively on "questions of socialism" and "proletarian culture."[49] During these years, in addition to all of his administrative, editorial, publishing, and teaching responsibilities, Bogdanov was also writing the third (and final) volume of his monumental treatise on tectology. Nor did he forget about blood exchanges. In the shortened "Outlines of Organizational Science," which appeared in installments in *Proletarian Culture* during 1919–20, Bogdanov included a brief section, entitled "The Tectology of the Struggle against Old Age," expanding on the ideas first aired in *Red Star* of using blood exchanges as a means to stave off aging. Bogdanov theorized that blood exchanges might represent a form of "physiological conjugation," somewhat similar to the conjugation of unicellular organisms. As the conjugation of infusoria "renews" their life to allow them to become virtually "immortal," Bogdanov (1920c, 30) surmised, so too would blood exchanges between human individuals allow them to "extend their life." But as far as I was able to ascertain, during these years, he never even thought of actually performing blood exchanges.

4 Earthly Realities

THE HEALTH OF THE RULING ELITE

Nothing in Bogdanov's writings up to 1921 indicates that he had seriously considered actually putting his vision of the "comradely exchange of life" and his idea of blood exchanges into practice. Certainly it was an exciting vision and an interesting idea, which had a place in his mind, but he was quite busy putting into practice his other, larger vision, that of proletarian culture, and especially its key (for him) element, proletarian science. Reality, however, intruded upon both his practical work and his vision.

A RIVAL AND A FRIEND

Perhaps Bogdanov would never have moved beyond his theoretical speculations to the practical applications of blood exchanges were it not for his old rival: Vladimir Lenin. Lenin was clearly suspicious of Bogdanov's popularity and authority within the Proletkult movement and among the countless "proletarian" students who attended Bogdanov's lectures and read his numerous books and articles. On December 1, 1920, at Lenin's prompting, the Central Committee of the Bolshevik Party denounced Proletkult as "an anti-Bolshevik organization" and placed it under the administrative control of Narkompros ("Pis'mo . . . ," 1920). Around the same time, the state publisher issued a new edition of Lenin's 1909 personal attack on Bogdanov and his ideas, *Materialism and Empiriocriticism,* with a print run of thirty thousand copies. In the short preface written in September 1920, Lenin admitted that he had had not changed his 1909 treatise, since he "had had no opportunity to familiarize himself with Bogdanov's latest works" (Lenin 1920, 10). To compensate for this "shortcoming," the new edition included a long article written by the well-known Bolshevik historian Vladimir Nevskii (1920). Characterizing Bogdanov's writings as "a philosophy of dead reaction," Nevskii updated Lenin's attack with a vicious critique of Bogdanov's recent publications (focusing in particular on the new, 1920 edition of his *Philosophy of Living Experience* and on the first two volumes of *Tectology*) written in the same caustic manner Lenin had used in his own assessment. The reissuing of Lenin's treatise ignited a wide campaign against Bogdanov in various party periodicals (Joravsky 1961).

But Lenin was clearly not satisfied. In late November 1921, when the Second Congress of Proletkult convened in Moscow, Lenin again pressed the party's highest council—the Politburo—into adopting yet another resolution targeting Proletkult as an instrument of "Menshevik" influence on the proletariat.[1] Accompanied by several articles written by Politburo member and leading party theoretician Nikolai Bukharin (1921a, 1921b) and published in the party mouthpiece *Pravda*, the attack prompted Bogdanov to resign from his post on the Proletkult governing council in order "not to taint the organization by his name."

The new attack on Proletkult also spurred fear among Bogdanov's friends that, as the movement's leading ideologue, he might be arrested or even executed. They had every reason to be fearful. At that very time, with the Red Army successfully extinguishing the last flames of the civil war along the borders of the new Soviet Russia, the Bolsheviks turned their attention to the "internal front" and began to liquidate all political opposition—both real and imaginary—to their "dictatorship of the proletariat." In late 1921, the secret police arrested, imprisoned, and often executed members of various "oppositionist" parties, ranging from anarchists and constitutional democrats to socialist-revolutionaries and Mensheviks. Indeed, during this very time, the secret police actively prepared a show trial of the leaders of the Socialist-Revolutionary Party—the country's most numerous and popular "agrarian" party and the Bolsheviks' major ally during the civil war years.[2] Even though the Proletkult stated goals centered on cultural not political work, this grassroots movement—by that time boasting over eighty thousand members around the country—presented a clear threat to the Bolshevik rule in the eyes of the secret police.[3] Bogdanov's old friends among high-ranking Bolsheviks were certainly aware that the Proletkult leadership might easily share the fate of other "oppositionists." One of them, Leonid Krasin (1870–1926), who together with Bogdanov had been expelled from the Central Committee of the Bolshevik Party in 1909 but now held the post of commissar of foreign trade in Lenin's government, took his old friend out of harm's way.[4] Apparently following the dictum "Out of sight, out of mind," Krasin invited Bogdanov to join him on the Soviet trade delegation to Britain as an expert in economics. Bogdanov readily agreed, and in early December 1921, he left for London.

Bogdanov's trip proved fateful indeed. After just a few days in London, he came across a newly released book entitled *Blood Transfusion*. The book was written by Geoffrey Keynes (1922), the younger brother of John Maynard Keynes—the prominent British economist whose works Bogdanov had carefully studied in preparation for his British mission[5]—perhaps piquing Bogdanov's curiosity even further. Geoffrey Keynes had served as a military surgeon in the Great War, and his book summarized his own experiences with blood transfusions as well as the latest advances in the technique made by U.S., Canadian, and British surgeons. According to his own account, Bogdanov immediately bought and hungrily read the book. He ordered apparatuses for blood transfusions and sera for blood typing from British suppliers. Perhaps he also used his time in London to peruse the newest literature on various subjects in experimental biology, particularly endocrinology and genetics, which—given the nearly total isolation of newborn Soviet Russia—was unavailable in Moscow.

Upon returning home, Bogdanov immersed himself in the preparation of a complete version of his *Tectology*. In the early fall of 1922, the Berlin-based publisher Zinovii Grzhebin—closely associated with Bogdanov's old friend Maxim Gorky—finally issued *Tectology*'s three volumes under one cover in a single massive tome (Bogdanov 1922). Busy with his editorial work, Bogdanov did not forget his London "discoveries": he enthusiastically began exploits in blood transfusions, with Keynes's book as his guide. Bogdanov's enthusiasm brought together a small circle of friends, including Semien Maloletkov, Bogdanov's fellow doctor during his stint in the army, and the physician-internist Ivan Sobolev, Bogdanov's boyhood friend at the Tula gymnasium. Sobolev invited into the circle his colleague from the Moscow University clinics, gynecologist Dmitrii Gudim-Levkovich.[6] Bogdanov's wife—an experienced nurse—also took part. They formed a study group that met regularly—usually once a month—in Maloletkov's apartment, discussing literature, learning the techniques of blood typing, and familiarizing themselves with blood transfusion equipment and procedures.

Bogdanov outlined the directions and goals of these exploits in a special paper he wrote sometime in late 1922 and distributed among his friends.[7] In surveying the current uses of blood transfusion and its technical intricacies, including blood groups and blood typing, the paper relied heavily on Keynes's book, which became Bogdanov's "transfusion bible." By far the largest portion of the paper, however, expanded on Bogdanov's earlier literary and theoretical excursions into the possible role of transfused blood in increasing the organism's "viability." Bogdanov reformulated his earlier ideas of the "comradely exchange of life" into a concept of "physiological collectivism," based on "breaking the boundaries of physiological individuality" and creating a new "physiological unity"—and thus increasing the "viability"—of numerous individuals through regular blood exchanges among them.

In the tradition of the clandestine operations he had engaged in during his days as a revolutionary conspirator, Bogdanov and his friends worked in secrecy. Bogdanov never tried to get financial, technical, or institutional support for his endeavors from Narkomzdrav, Narkompros, or any other state agency, covering all the considerable expenses of his circle out of his own pocket.[8] But Bogdanov's regular meetings with his "brothers-in-blood-exchanges" did not escape the notice of one state agency: on the night of September 8, 1923, the secret police searched Maloletkov's and Bogdanov's apartments. Bogdanov was arrested and confined to the infamous Lubianka prison. Bogdanov suspected that his arrest was somehow linked to the "organization of physiological collectivism," as he named his circle.[9] But as he found out after a week of imprisonment, he was accused of leading an oppositionist group, "Workers' Truth," that had actively used Bogdanov's ideas in their critique of the Bolsheviks.[10] After two weeks of interrogations by secret police investigators, during which Bogdanov had repeatedly reiterated his absolute disinterest in any sort of active political work, he wrote an indignant letter to the head of the secret police, Felix Dzerzhinskii, an old Bolshevik whom Bogdanov had known during his days in the party, demanding a face-to-face meeting. Dzerzhinskii did meet with Bogdanov, and their conversation seemingly cleared up all "misunderstandings." Bogdanov even arranged for the delivery to Dzerzhinskii of a sample of his recent publications to dispel whatever doubts the head of the secret police might still be harboring regarding Bogdanov's views on and involvement with contemporary political struggles. But despite repeated appeals from Bogdanov's friends among high-ranking Bolsheviks, he was released from the prison only two weeks later, on October 13.

The "incident" proved seriously injurious to Bogdanov's health: he nearly suffered a heart attack awaiting his release. It possibly played a role in Bogdanov's decision to move from theorizing on the questions of physiological collectivism to attempting its practical implementation. Or perhaps, the death of his old rival, Lenin, on January 21, 1924, prompted Bogdanov to begin experimenting with blood exchanges. Only three years younger than Lenin, Bogdanov had undoubtedly been aware of Lenin's prolonged illness that led to Lenin's premature death, and perhaps it heightened Bogdanov's sense of his own mortality. Whatever the case, Bogdanov and his friends did not even consider performing preliminary tests on animals: in the best tradition of medical pioneers, Bogdanov tried the procedure first on himself.[11]

On February 11, 1924—the very day the Communist Academy held a special meeting commemorating Lenin's death—Bogdanov attempted a "blood exchange" for the first time.[12] His "companion," as Bogdanov called him, was a twenty-year-old male student. Due to some technical problems, the exchange was unsuccessful, and the experiment ended with a simple transfusion of 330 cubic centimeters of blood from the student to Bogdanov. A week later, Bogdanov tried the procedure for a second time. But again an exchange proved impossible, and the attempt ended with Bogdanov receiving another 340 cubic centimeters of blood from his "companion." Three months later, on May 18, 1924, the exchange was tried once again. Finally, it worked: Bogdanov received 700 cubic centimeters and the student 500. According to Bogdanov, in the course of the subsequent months—both after the first two transfusions and after the final exchange—his health improved dramatically. Not only did he feel much better subjectively, but several objective measurements, including blood cell count and blood pressure, pulse frequency, muscle tone and strength, as well as lung capacity, showed marked improvements.

In November 1924, two more members of the group, Bogdanov's wife, Nataliia Korsak, and his collaborator, Semien Maloletkov, took part in blood exchanges with a female and a male student, respectively, with "no negative results." Thirteen months after the first failed attempt, on March 29, 1925, Bogdanov repeated the blood exchange with another "companion"—a twenty-three-year-old male student. This time Bogdanov received 800 cubic centimeters of blood and the student 830. The new exchange was apparently prompted by a significant deterioration of Bogdanov's health: the improvements spurred by the previous ones had largely disappeared. A few months after the exchange, Bogdanov again felt noticeably better; and, according to a witness, he "looked great; obviously, the operation had helped him considerably."[13] On May 10, 1925, an exchange was performed between another pair: a forty-two-year-old female doctor and a twenty-eight-year-old female student. The doctor received 375 cubic centimeters of blood, the student 750. On November 4, 1925, Bogdanov participated in a blood exchange with yet another "companion"—a thirty-seven-year-old male "statistician." According to a witness:

He himself and [his wife] Nataliia Bogdanovna look great, and, I think, he has become younger if not by ten, then by seven or five years, for sure. . . .

A recent photograph [*sic*] shows that even the diameter of his aorta has decreased! A completely incredible thing, but it's a fact. Furthermore, it corresponds completely to his subjective feelings [*samochuvstvie*]: forgetting [his age and health], sometimes he runs up four to five flights of stairs. Nataliia Bogdanovna also feels great: gout symptoms in her feet are gone; before she had to wear shoes made to order, now she wears regular ones.[14]

A few weeks later, on November 29, 1925, Maloletkov exchanged first 800 cubic centimeters of blood, and then 650 more, with another twenty-nine-year-old male student.

During nearly two years of experimentation with blood exchanges, neither Bogdanov himself nor any other member of his circle published a single report on their techniques, experiments, and findings. Only a chosen few knew of the existence of Bogdanov's group and its work. Among these selected few was Leonid Krasin.

As it happened, in November 1925, Krasin fell seriously ill.[15] Blood tests performed at the Kremlin clinic indicated that he had developed acute anemia: both his level of hemoglobin and his red blood cell count were considerably lower than normal. Two Kremlin doctors, the leading Russian internists Dmitrii Pletnev and Vasilii Shervinskii, recommended that Krasin take a prolonged rest somewhere in a warm and sunny climate, but could neither determine the causes of his illness nor offer any reliable therapeutical treatment. They suggested that Krasin travel to Berlin to consult with German specialists. Instead, a few days after he had been released from the Kremlin clinic on November 22,[16] Krasin went to see his old friend Bogdanov, with the idea that a blood transfusion might alleviate his condition.

Initially, Bogdanov was not very enthusiastic about Krasin's idea. He only recommended that instead of Berlin, Krasin travel to Paris or London, where "since the Great War, the science of blood had advanced considerably, while the Germans had lagged far behind." In a few days, however, Bogdanov called Krasin on the phone and invited him for a visit. As it turned out, Krasin's idea prompted Bogdanov to review the available literature, and sure enough, in his "transfusion bible," he found the medical histories of a number of patients with anemia similar to Krasin's. According to Keynes, in 60 percent of the cases, a blood transfusion completely cured the patients. Bogdanov showed Krasin the book and offered to transfuse 700–800 milliliters of blood. Even if the transfusion did not cure Krasin's illness, Bogdanov asserted, it would give him enough hemoglobin and red blood cells to "strengthen his organism" and thus allow him to undertake a trip abroad to continue treatments. Krasin agreed, and Bogdanov began searching for a donor. A friend of Krasin's offered his blood but was rejected because of the incompatibility of their blood groups. In a few days, a suitable donor was found, and Krasin received his transfusion. Reportedly, the procedure improved Krasin's condition dramatically. A few weeks later, in early 1926, he left Moscow for a vacation in southern France before taking up the post of Soviet ambassador to Britain.[17]

But before Krasin left Moscow, word of his "miraculous" recovery had apparently spread. It found an attentive ear in the general secretary of the Bolshevik Party, a member of its Politburo, Joseph Stalin. One day in late December 1925, according to Bogdanov, he was called to the Kremlin for a meeting with Stalin.[18]

THE PATRON

Why, we may wonder, did Stalin take an interest in Bogdanov's work on blood exchanges? Some commentators have alleged that Bogdanov's ideas appealed to the Marxist in Stalin, since they resonated with his own ideological convictions.[19] It may well be that Stalin shared Bogdanov's version of Marxism (e.g., Alekseev 2006), but it seems more probable that Stalin had much more immediate concerns than the philosophical/ideological implications of Bogdanov's experiments. In 1925, Stalin was quite far from being an omnipotent ruler, without whose approval nothing could take place in the country. In fact, at the time, his power was rather limited: he was not even a member of the country's highest governing body—SNK. But, as the general secretary, he was in charge of party cadres. And it was exactly this responsibility that most likely prompted Stalin's interest in Bogdanov's work, for in 1924–25, a virtual "epidemic of deaths" plagued the Bolshevik Party.

In the immediate post–civil war years, the Bolshevik Party lost a number of its prominent members to what party doctors termed "revolutionary exhaustion and attrition" (Zalkind 1925). Commissar Semashko (1923) even identified "nervous disorders and overexhaustion" as "professional diseases of communists." Indeed, according to the party's internal reports, nearly half (44 percent) of all visits to medical establishments by top-level party members were due to "nervous disorders," with tuberculosis holding a distant second place at 27 percent and all intestinal, cardiac, and other internal diseases combined accounting for only 29 percent of visits.[20] Many of these "disorders" apparently proved fatal. The country's central newspapers, *Pravda* and *Izvestiia,* carried on a regular basis the obituaries of high-ranking party and state bureaucrats, who, in the Soviet parlance of the day, "had burned out on the job."

Of course, the Bolsheviks had always taken special care of the health of their leaders. After the move from Petrograd to Moscow in the spring of 1918, the Bolshevik government and party apparatus settled in the Kremlin, a large, walled-off complex of buildings in the center of Moscow, which for centuries had served as a residence to the Russian monarchs. In the fall of 1918, an epidemic of typhus engulfed the newborn Soviet republic. In October, to provide medical care to the Kremlin inhabitants who fell victims to the scourge, a small clinic with ten beds and an outpatient unit with several doctors' offices were set up in one of the Kremlin's numerous buildings. The epidemic reached its peak during the next winter, prompting Lenin to issue his famous call to arms: "Either the lice will defeat socialism, or socialism will defeat the lice." Obviously, socialism could not survive without its leaders, and in February 1919 SNK created a "Kremlin Sanitary Board" specifically to prevent the spread of the epidemic to the government compound.[21] The board immediately set up "sanitary cordons" (not too many, thanks to the high walls and very few entrances to the Kremlin) to check for signs of disease in everyone entering the Kremlin and to delouse and disinfect their clothing. It also organized public baths, laundry facilities, and a garbage incinerator. At the same time, to limit contacts with the sick, the board established a large hospital specifically for typhus patients outside the Kremlin walls. In April, given the total lack of medical supplies in the city, the board attached to the Kremlin clinic a small but well-stocked pharmacy funded directly through SNK.

With the end of the civil war, in early 1922, the board was reorganized and expanded. All institutions providing health care to the Kremlin inhabitants came under the board's purview, including the Kremlin clinic (with its outpatient units, dental offices, and pharmacy), various dachas near Moscow, and several health resorts in the south of the country. The Kremlin clinic grew to fifty beds and established its own "analytical laboratory" to conduct necessary tests on the premises. It also acquired a "professorial staff" that included three well-known physicians: Fedor Get'e, the head of Moscow's largest hospital; Vladimir Shchurovskii, former personal physician to Leo Tolstoy; and Dmitrii Pretnev, a rising star internist from the Moscow University clinics. As needed, the "professors" invited other specialists, from an ophthalmologist to a dietician, for consultations. The board also arranged for consultations by eminent German physicians invited to come to Moscow specifically for that purpose.

The Kremlin Sanitary Board was subordinate to SNK and served the medical needs of all Kremlin inhabitants, but the party apparatus found this arrangement insufficient. In October 1923, the Central Committee of

4.1 Lenin's doctors (*left to right*): Aleksei Kozhevnikov, Oscar Minkowski, Nikolai Semashko, Otfrid Foerster, and P. I. Elistratov, circa 1923. (Courtesy of the Museum of the History of Medicine of the Sechenov Moscow Medical Academy.)

the Bolshevik Party instituted its own "Medical Commission" (*Lechebnaia komissiia*) headed by S. I. Filler, a member of the party's internal watchdog agency, the Central Control Commission (TsKK).[22] Filler's commission administered a huge fund of nearly one million rubles to address the health needs of top-level party members, including treatments in various medical institutions within the USSR, treatments abroad, and cash payments for treatments "outside of medical establishments."[23] In a single year of its operations, from October 1923 to October 1924, Filler's commission served nearly thirty-five hundred party officials, arranging and paying for their placement in hospitals, sanatoria, and health resorts in various parts of the country. The commission also spent almost one hundred thousand rubles on fifteen "patients" sent abroad—to Germany, France, Italy, and England—for health reasons.

The highly publicized death of Bolshevik Party leader Lenin in January 1924 prodded the government apparatus to create—under the SNK auspices—a "Supreme Medical Commission," headed by Aleksei Rykov, Lenin's successor as the head of SNK, with Commissar Semashko and Viktor Radus-Zen'kovich, a Bolshevik doctor and TsKK representative, as members.[24] SNK provided the "supreme" commission—referred to in internal correspondence simply as "the troika"—with a special sizable fund (of 150,000 rubles) to be spent on the health needs of top-level government officials, including trips abroad for medical consultations, treatments, and vacations.

The party apparatus followed suit. On January 31, 1924, just ten days after Lenin's death, a plenary meeting of the Central Committee adopted a special resolution on the "protection of health of the 'old party guard.'" The resolution charged TsKK with the task of elaborating "necessary measures for the protection of health of the party's elite [*partverkhushka*]."[25] In mid-June, the Politburo listened to a report prepared jointly by Narkomzdrav and TsKK. Obviously dissatisfied with what they had heard, the Politburo instructed "Comrades Semashko and Radus-Zen'kovich to develop measures for the protection of health of the old party guard and present it to the Politburo for final approval in one week."[26]

But all these commissions, resolutions, orders, and instructions seemed to produce little effect: the "epidemic of deaths" among high-level Bolshevik officials continued. On October 31, 1925, Mikhail Frunze—the legendary forty-year-old commander in chief of the Red Army, commissar of the army and the navy, and head of the country's Revolutionary-Military Council—died during a surgical operation.[27] Frunze had had a long history of stomach bleeding due to a perforated ulcer, which had previously been treated with the only method then available—a surgical excision. The operation helped, and for several years Frunze felt much better. But in late 1924, his condition worsened. He went abroad for two months of treatments with mineral waters. In the summer of 1925, Frunze also took a prolonged rest at a health resort in the Crimea. But nothing seemed to help. In the fall, upon Frunze's return to Moscow, twelve (!) leading internists and surgeons were summoned to the Kremlin clinic for a consultation. They decided that Frunze had again developed a perforated ulcer and suggested that a new operation be performed at one of the city's hospitals.[28] It was supposed to be routine, but, according to the official autopsy report published in *Pravda,* Frunze died during the operation of a heart failure provoked by general anesthesia.[29]

Frunze's death shocked the country and shook its leaders.[30] Within two weeks, Rykov's commission prepared a special report on "the protection of health of the leading [*otvetstvennye*] workers of the party and the government of the USSR."[31] Presented by Rykov to a Politburo session on November 19, the report suggested that the Kremlin clinic be expanded to include its own surgical and therapeutic wards and thus ensure the "individualized treatment" of its patients. The report also proposed that two eminent German physicians—Friedrich Kraus, a prominent cardiologist, director of the Charité Hospital in Berlin, and

Otfrid Foerster, a famous neurologist from Breslau who had treated Lenin during his illness in 1922–23—be invited to Moscow for a thorough examination of the health of Soviet leaders. The commission compiled "List No. 1," naming fifty top-level members of the party and the government who required "particular attention to their health" and were to be examined by the invited German specialists. The Politburo approved Rykov's proposals.[32]

But in a few days, they had to revisit the issue. The reason was a long letter Leon Trotsky, a member of the Politburo, sent to his fellow Politburo members.[33] Trotsky did not attend the November 19 meeting: he himself was recovering from a long illness in Kislovodsk, a health resort in southern Russia renowned for its mineral waters. But he did receive the meeting's protocol and found Rykov's proposals insufficient, if not completely wrongheaded. "Frunze's surgical operation and his death," Trotsky declared, bore witness to the "unsatisfactory organization of the system of medical monitoring of [the health of] leading workers." It should not have happened, Trotsky exclaimed, that Frunze's surgeons had not known the simple fact that he had a weak heart and thus could not undergo general anesthesia. Trotsky asserted that the system of occasional medical consultations was not simply inefficient but downright harmful. In his opinion, such consultations brought together "people who do not know the patient's organism, and who are forced to come, in an hour or two, to a joint decision. The main worry in such cases—is not to offend each other. Every decision is made as a compromise among different medical points of view and different medical ambitions. And this compromise is made on the patient's back. Every serious and honest physician will admit it."

Instead of occasional consultations, Trotsky proposed instituting a system of continuous monitoring. Every leader should be assigned one physician responsible for his patient's health. The physician should keep a "health diary" supplemented with "anatomical and functional passports" of the patient's particular organs (heart, liver, lungs, kidneys, etc.). In this way, in case of an acute illness, like Frunze's stomach ulcer, the "responsible physician" would be able to give attending surgeons or specialists all the necessary information on the health status of their patient. "This information," Trotsky insisted, "would be much more important and valuable for making an appropriate decision than an inconsequential consultation, during which doctors make life-and-death decisions based on hearsay or on superficial observations and questioning."

At the beginning of December, Trotsky returned to Moscow and on December 3, he voiced his concerns at a session of the Politburo. The

Politburo charged Rykov's commission with the task of developing a comprehensive plan to reform the existing system of medical services for the country's ruling elite.[34] In a few days, Pavel Obrosov, the head of the Kremlin clinics, prepared a twenty-five-page report "On the Reorganization of the Kremlin Clinics," which incorporated Trotsky's proposals on health monitoring—including the assignment of "responsible physicians" and the creation of "health passports" of their patients—into the existing plans of the clinics' considerable expansion. Obrosov's report landed on Stalin's desk.[35]

It was during this very time, permeated by anxieties about the health of the Bolshevik leadership and heightened by Frunze's sudden death, that Krasin—number 36 on "List No. 1"—fell ill and was "cured" by a blood transfusion performed by Bogdanov's "organization of physiological collectivism." And it was against this backdrop that Stalin invited Bogdanov to the Kremlin for a meeting in late December 1925.

We can only speculate on the actual details of their conversation: there are no records of the meeting in the party archives; Bogdanov's appointment was not even logged in Stalin's "visitors' journal."[36] But it seems logical to suggest that one subject was Frunze's death. By a strange coincidence, just one week before Frunze's fateful operation, on October 23, 1925, *Izvestiia* carried a short article about a similar operation, performed by the well-known Leningrad surgeon Erik Gesse on a patient with a perforated stomach ulcer ("Perelivanie krovi" 1925a). One of the most vocal advocates of blood transfusions, during the operation, Gesse transfused his patient with blood. The newspaper reported that "the patient had fully recovered" after the operation and since "gained forty-one pounds," implying that it was the blood transfusion that secured the operation's success. If Stalin had read the article—and by all available evidence, he was an attentive reader of *Pravda* and *Izvestiia* (often before the newspapers were released for sale on the street)—it might have given him enough "food for thought" and for a pointed discussion of blood transfusions with Bogdanov. Or perhaps they talked about Krasin's "miraculous" recovery after the blood transfusion performed by Bogdanov's circle. Regardless of what exactly Stalin and Bogdanov discussed, their meeting appears to have had no immediate consequences.

Some six weeks later, on February 11, 1926, Rykov and Semashko prepared for the Politburo a draft resolution "on the protection of health of the leading workers of the party and the government."[37] The draft combined Trotsky's proposal of continuous health monitoring with frequent consultations by prominent physicians, including regular—semiannual—consultations by "foreign specialists": the hefty sum of twelve thousand U.S. dollars per annum was allocated for that purpose. The draft also incorporated suggestions presented by Rykov's commission in November and elaborated in Obrosov's report to Stalin of expanding the Kremlin clinics to create "separate therapeutic, nervous, and surgical wards" along with "an exemplary prophylactorium." The resolution charged two top-level members of the party apparatus, Stanislav Kosior (a secretary of the Central Committee) and Valerian Kuibyshev (the head of TsKK), with overseeing its implementation.

A week later, on February 18, after Rykov's and Filler's commissions had hammered out the details, dividing up allocated funds and responsibilities, the Politburo was supposed to approve the resolution.[38] But the decision was postponed: the Politburo had a more pressing issue on its agenda. That very morning Friedrich Kraus and Otfrid Foerster arrived in Moscow, and the Politburo had to make practical arrangements for medical consultations by the two "foreign specialists." The consultants had come to Moscow for only eight days and faced a considerable challenge: to conduct thorough physical checkups of all individuals included on "List No. 1," which of course required a lot of preparation on the part of party managers and gofers. The Politburo approved the schedule of the consultations and ordered all the "comrades included on the list to free themselves for the entire day"; each of them was assigned in the schedule. "Attendance is obligatory, unconditionally," the Politburo's order stated sternly.[39]

The "comrades" obviously followed the orders, and the Germans accomplished their task admirably. Beginning on the day of their arrival, Thursday, February 18, they saw their first two "patients"—the head of the secret police, Feliks Dzerzhinskii, and his deputy Viacheslav Menzhinskii. For the next seven days, they examined five to six "patients" daily, taking off only one day—Sunday—for a meeting with their Soviet colleagues. On Friday, February 26, Kraus and Foerster saw the last group, which included Stalin and Rykov. In the evening, their mission accomplished, the German professors left Moscow, splitting a generous honorarium of five thousand U.S. dollars and leaving behind "List No. 1" with a new registry that recorded the diagnoses of major health problems of each "leading worker" confirmed by the consultants.[40]

Perhaps it was a simple coincidence, but on that very day, the man responsible for the implementation of the February Politburo resolution, Valerian Kuibyshev, sent a note to one of the highest government agencies—the Council of Labor and Defense (STO), where he served as a deputy-chairman.[41] The note informed the council that Kuibyshev had approved a decree on the organization of an institute of blood transfusion and had issued an eighty-thousand-ruble advance to Narkomzdrav for its establishment. The note also instructed the People's Commissariat of Finance (Narkomfin) to approve the institute's budget for fiscal year 1925–26 within a month's time.

Kuibyshev's note made its way through the intricate system of Soviet government agencies with incredible speed. As soon as it was received, and without any discussion, the council rubber-stamped Kuibyshev's instructions, issued them as its own resolution, and forwarded it to the SNK of the Russian Federation, which in turn issued its own decree to the same effect on the same day.[42] The next day, Semashko signed his own order establishing the institute under the auspices of Narkomzdrav, and the day after that, he appointed Bogdanov its director. The "world's first" Institute of Blood Transfusion came into being, at least on paper. Its newly appointed director now faced the challenging task of making it real.

THE INSTITUTE

The creation of a scientific institution involves solving a variety of problems, ranging from finding a suitable building, devising an appropriate internal structure, and hiring experienced personnel to producing the floor plans for separate research and clinical departments and obtaining necessary supplies and equipment. Bogdanov attended to this formidable task with panache. At the beginning of March 1926, a government commission in charge of the city's space allocation assigned the institute a large building in the center of Moscow, just one mile from the Kremlin. A prominent city landmark—a mansion of the wealthy merchant Nikolai Igumnov built in the 1890s in a "pseudo-Russian" architectural style—the building at the time housed a "cultural club" for the workers of a nearby factory.[43] Bogdanov immediately sealed the building and changed the locks. Despite the vocal protests of the factory's trade union to various governmental agencies, on March 23, SNK affirmed Bogdanov's mandate.[44] With the eighty-thousand-ruble advance issued by Kuibyshev in hand, plus sixty thousand more rubles allotted by Narkomfin for the institute's annual budget, Bogdanov began to plan the renovations needed to convert the building into a research institute, to order equipment and materials, and to hire personnel. The question of hiring appropriate personnel is obviously one of the most important in creating a research institution. Bogdanov did not resort to the customary system of announcing available vacancies and selecting the best among the applicants. Instead, he hired his "comrades-in-blood-exchanges." His "organization of physiological collectivism" formed the core staff of the new institute, with Maloletkov as "deputy

4.2 The Igumnov mansion

director for scientific and educational affairs," Sobolev as head of the "therapeutic section," and Gudim-Levkovich as head of the "surgical section." Semashko also appointed a trusted party member, D. L. Zeilidzon, as "deputy director for administrative affairs" to assist Bogdanov with numerous technical issues.

Around the same time, without the customary discussion by the Narkomzdrav Scientific Medical Council, Semashko personally approved "a temporary statute" for the institute, giving it a very special status.[45] Unlike other research institutions under Narkomzdrav's patronage, the Institute of Blood Transfusion was not created as a part of GINZ or the Narkomzdrav Clinical Division. Instead, it was subordinate to the commissariat's Administrative-Financial Directorate. This meant that Bogdanov had to report directly and solely to the commissar himself. Very few institutions among Narkomzdrav's impressive research empire shared the status accorded to Bogdanov's institute: one was the laboratory created in late 1925 to study Lenin's brain, another the German-Soviet laboratory for racial pathology created in 1927.[46] It is also quite telling that all three institutions were eventually housed in the Igumnov mansion.

In addition to locating the new institute within Narkomzdrav's complex institutional structure, the statute also enumerated its tasks. First on the list was "to study and elaborate issues related to blood transfusions." Second place was given to "the theoretical and practical instruction of physicians through the organization of occasional and permanent courses on blood transfusions." The third task was "publication of scientific and popular literature on blood transfusions." Last on the list was "the manufacturing of standard sera, as well as preparations, apparatuses, and accessories for blood transfusions."

Bogdanov attended to his assigned tasks immediately, but he also had his own priorities. Just a few days later, on April 4, 1926, he published a long article on the aims and prospects of his institute in *Izvestiia*, the country's most widely distributed newspaper, whose editor in chief, Ivan Skvortsov-Stepanov, was Bogdanov's old friend and coauthor. After a lengthy excursion into the history of blood transfusions, the discovery of blood groups, and the Great War experiences with the procedure, Bogdanov (1926) surveyed the "latest foreign literature" on the subject. He asserted that, following in the footsteps of their "perpetual teachers"—German physicians—Russian doctors had lagged, "and criminally so," far behind their French, British, and, particularly, U.S. colleagues in adopting this "lifesaving" operation.

He described his own long-standing interest in the procedure, which had led him to formulate the concept of "physiological collectivism"— the increase of the "viability" of individual organisms through regular blood exchanges among them. Bogdanov reported that in early 1924, he had organized a research group to study "physiological collectivism." Demonstrating his complete ignorance of (or perhaps a profound disdain for) the available Russian literature, Bogdanov claimed that his group was the first in the country to practice "the modern technique of blood transfusion" based on matching the blood groups of the donor and the recipient. He mapped out the current fields of application for blood transfusions: war wounds and trauma, anemia, carbon monoxide poisoning, surgical operations that involved excessive bleeding, septicemia, and burns, which naturally brought him to his own ideas about blood exchanges and his own experiments.

Bogdanov reported that to date his group had performed "about ten" blood exchanges. In all cases but one, nearly one-seventh of the patients' entire blood volume had been exchanged; in one case—more than one-fourth. The results—because of the small number of cases— "do not allow [me] to draw decisive conclusions," Bogdanov admitted, "but they are certainly encouraging and call for the continuation of research." In seven cases, the exchange had been performed between an old and a young individual; and in "the majority of cases (five out of seven), both persons have benefited," in the exact manner predicted by Bogdanov's theory of "physiological collectivism."

"Whatever would come out of our special line of investigations," Bogdanov continued, "even in the framework of current Western research, blood transfusions have enormous scientific and social import."

It was precisely because he understood both the "social-practical and scientific importance" of blood transfusions, Bogdanov claimed, that Narkomzdrav had assigned him the task of organizing the institute. The country needed a special institute for blood transfusions, Bogdanov asserted, to ensure that this important procedure was available not only to the Red Army but also to the "labor army," as well as to the "state apparatus, whose best workers are wearing out far too quickly."

According to Bogdanov, the creation of such an institute "will not be easy: even an examination of Western experiences and the transfer of their results to our country is no mean feat." The institute would create "the simplest and most suitable to our conditions type of apparatus, together with a detailed and sufficiently comprehensive manual for blood transfusion," Bogdanov promised. "We will inform Russian physicians of the current state of affairs on the basis of the best available [foreign] monographs, taking one of them as the foundation [for the manual]." To further this goal, he concluded, the institute "will send its specialists abroad to learn at the source what has been done and how it is done."

A few weeks later, an announcement in the *Moscow Medical Journal* affirmed that the institute's major purposes were "research and propaganda," not clinical (*lechebnoe*) application of blood transfusions. The journal repeated Bogdanov's claims: "In America and Europe, the operation of blood transfusion is made annually in tens of thousands of cases. The Soviet Union has lagged behind and must catch up as soon as possible."[47]

The public announcements of the institute's broad goals and prospects notwithstanding, even before the renovations were under way, the institute was called upon to fulfill its main, though unstated, duty: to take care of the health of "leading workers." "Soviet exhaustion," as the Kremlin doctors had termed it, was taking a heavy toll on the health of party and state bureaucrats. In May, the institute staff transfused blood to "one of the leading workers of the party" who suffered from acute anemia. At the end of June, another "leading worker" with a case of "nervous overexhaustion" received a blood transfusion. On July 1, the institute personnel treated with a transfusion yet another "leading worker from the provinces."[48]

By the end of October 1926, after nearly seven months of continuous struggle with shortages, bureaucratic hurdles, and administrative incompetence, the renovations of the Igumnov mansion were completed. Although its operating room, laboratory for blood typing, and ten-bed clinic still lacked some equipment, the institute was ready to pursue its stated goals. Alas, Soviet physicians waited in vain for "the simplest apparatus" or the courses on blood transfusions, which Bogdanov had promised them in April. Instead, in December, Bogdanov (1927a) published a 160-page treatise, eloquently entitled *Struggle for Viability*.[49] The book came out under the auspices of the Institute of Blood Transfusion, with a print run of five thousand copies. But it was not the "comprehensive manual" on blood transfusions filled with information on "the current state of affairs on the basis of the best available monographs" that Bogdanov had promised his readers either. In actuality, the book was an expanded and updated version of Bogdanov's "tectological" theory of senescence.[50]

5 Struggles for Viability
PROLETARIAN SCIENCE IN ACTION

Obviously, Bogdanov had no doubts regarding his own qualifications for and ability to direct the "world's first" institute of blood transfusions, nor did he have any qualms about accepting an appointment as its director. Yet this appointment brought Bogdanov's private exploits in blood exchanges onto the public stage and, more important, thrust them into the existing intellectual and institutional contexts of Soviet transfusiology, and of biomedical science more generally. The support of the Bolshevik Party's leadership—most notably Stalin and Semashko—secured Bogdanov's institutional position. But his intellectual position appeared much more tenuous. During the preceding years, Bogdanov had had to persuade only his closest friends and collaborators of the "viability" of his ideas and techniques. Now he needed to justify them to a much broader audience. This audience consisted not only of the general public, which could be swayed by simple newspaper "advertisements" such as Bogdanov's April 1926 article in *Izvestiia*. It also included the professional medical and scientific communities, which obviously required something more substantial than a newspaper article to be convinced of the validity and "viability" of both Bogdanov's ideas and his institute. This was probably the main reason Bogdanov spent the summer and fall of 1926 writing his *Struggle for Viability*.

The book clearly presented an exercise in applying the basic principles of proletarian science Bogdanov had been elaborating for nearly two decades in both fictional and philosophical writings to his current task at hand: studying blood transfusions. In his numerous philippics, Bogdanov had lambasted *specialization*—along with its main instrument, specialized, esoteric language—as the defining features of individualistic "bourgeois" science, which made it incomprehensible to the proletariat and thus transformed science into yet another means of perpetuating the hegemony of the bourgeoisie in a capitalist society. By contrast, the defining features of collectivist proletarian science, according to Bogdanov, were its *universality* and its generalized and simplified language, which made it understandable to anyone with some basic education. Accordingly, Bogdanov emphasized in the preface: "The book is written in such a way that it could be read not just by physicians but by anyone who has some knowledge of natural sciences." "Following the plan developed by the institute of blood transfusion," Bogdanov continued, his book was to serve as "a general introduction to a series of other works, both original and translated, which would be devoted to the elaboration of the question of blood transfusion from the medicopractical side" (Bogdanov 1927a, 3).

For Bogdanov, his *Tectology* constituted the pinnacle of "universal" proletarian science. Thus, it is hardly surprising that Bogdanov employed his "tectological" categories to address the issues of blood transfusions in the most generalized form, as he put it himself, "largely from the general biological point of view, and only partially from the proper medical viewpoint." Indeed, he incorporated the corresponding paragraphs from *Tectology* into his new text, practically without changes. But he did not produce a "general theory" of blood transfusions, as one might have expected. Instead, nearly three-quarters of his book dealt with the elaboration and justification of a "tectological" theory of senescence.

Undoubtedly, Bogdanov had serious intellectual reasons for adopting this particular framework for his ideas about blood exchanges. Even though in *Red Star,* the Martian doctor Netti asserted that blood exchanges "rejuvenate" their participants only to a certain extent, issues of aging and rejuvenation provided a major background for further development of Bogdanov's ideas. In his 1910 program, in his 1913 musings in the first volume of *Tectology,* in his 1920 outline of its final volume, and in his 1922 paper on "physiological collectivism," Bogdanov raised again and again the possibility that blood exchanges could indeed stave off the aging processes. So, following the prescriptions of his proletarian science, he needed first to "generalize" his views on the subject of senescence in order to explain how exactly blood exchanges could—and apparently did in his own experiments—affect aging.

AGING AND REJUVENATION

Bogdanov's formulation of the general framework for his ideas was influenced not merely by his previous intellectual pursuits. Although the issues of aging and dreams of rejuvenation had dogged humanity for centuries, in no other time did these issues and dreams generate such incredible public appeal and fascination as in the 1920s.[1]

The rise of experimental biology and medicine in the last decades of the nineteenth century had inaugurated a new era in the study of senescence. Armed with new experimental methods, scientists and physicians attacked the centuries-old mystery of aging, and it seemed they would quickly succeed in finding suitable techniques to control and even reverse the aging processes. In 1889, Charles Brown-Sequard, a prominent French physiologist who had succeeded the "father" of experimental medicine, Claude Bernard, at the College de France, astonished the world with the announcement that injections of extracts prepared from animal testicles had "rejuvenating" effects on the human organism: he had tried it on himself, and it had certainly worked. Brown-Sequard's basic idea was an old one—namely, that aging results from the decline in the activity of sex gonads. But this old idea received a new interpretation in Brown-Sequard's work: aging resulted from the insufficiency of the "internal secretions"—a concept developed by Claude Bernard—produced by the gonads.[2] Hence, with the new experimental methods, the aging process could be reverted by supplying the aged organism with the secretions (extracts) of young sex glands. Although initially greeted with much skepticism by the medical community of the time, Brown-Sequard's research created a new direction in therapeutics—organotherapy, treatments with various animal tissue extracts—and paved the way for the discovery of hormones and the emergence of a new science of "internal secretions," endocrinology.[3]

Another branch of the new, experimental medicine—bacteriology— also promised a solution to the problems of senescence. In the early 1900s, Elie Metchnikoff, a Russian zoologist working at the Institut Pasteur in Paris, published his theory that attributed aging to the "poisoning" of organism with toxins produced by bacteria inhabiting its intestines. Metchnikoff (1901–2, 1903) suggested that inhibiting the growth of "noxious" bacteria in the intestines by replacing them with "good" bacteria (such as the milk-souring bacteria present in yogurt) would extend human life considerably.[4] One of the founders of the new science of immunology, Metchnikoff also advocated for the establishment of a special science of senescence and even coined a name for it: gerontology.[5]

Brown-Sequard's and Metchnikoff's work helped establish senescence as a legitimate subject of scientific inquiry and inspired other scientists to begin their own investigations. During the early 1900s, researchers working in two branches of experimental biology, cytology and biochemistry, offered yet another view of senescence as the result of progressive accumulation in the cells, particularly nervous cells, of certain "waste products." A number of researchers described changes in the appearance and chemical composition of "old" cells as compared to "young" ones. Some of them theorized that if there were a way to remove these "toxic wastes" from cells, the organism could be "regenerated" and its aging would be delayed dramatically. Experiments on cultures of infusoria (and later on tissue cultures) kept in the laboratory seemed to confirm this hypothesis: the cultures whose nutrient medium was regularly "renewed" lived much longer (if not indefinitely) than the cultures whose medium was allowed to accumulate their "waste products."[6]

Since Brown-Sequard and Metchnikoff came up with seemingly simple "remedies" for aging—injections of testicular extracts and yogurt drinking—their concepts attracted much public attention and raised public expectations that, quite soon, science would be able to fend off old age, if not indefinitely then at least for a very long time. But it was the works of the Austrian physiologist Eugene Steinach and the French surgeon Serge Voronoff that in the early 1920s spurred an extraordinary obsession with "rejuvenation" around the globe, including Russia.[7]

In June 1920, Steinach, a respected member of the famous Viennese Institute of Experimental Biology (the "Vivarium"), published a seventy-page article in the Berlin-based journal *Roux's Archive of Developmental Mechanics*. Entitled "Rejuvenation through the Experimental Reinvigoration of the Aged Puberty Gland," the article continued Steinach's investigations begun almost fifteen years earlier into the role that the internal secretion of sex gonads—testicles and ovaries—plays in the animal organism. These investigations had led Steinach to the idea that sex gonads consisted of two anatomically and histologically distinct parts. One was responsible for the production of sex cells (ova and sperm), the other for the production of sex hormones. He named the second part the "puberty gland." The 1920 article detailed Steinach's newest experiments with ligation of seminal vesicles (vasoligature—an operation similar to vasectomy) in old male rats, as well as with transplantation of sex glands (both male and female) from young rats to old ones. Steinach claimed that both vasoligature and transplantation stimulated the "puberty gland" and produced clear rejuvenating effects in operated animals. In a space of a few weeks following the operation, a weak, bald, and sexually impotent rat was transformed into one that was strong, covered with new fur, and sexually aggressive. The operated animals also lived much longer than the controls; the operations extended the rat's life span by a third, from thirty months up to forty months. Supplemented by several photographs of rats before and after "rejuvenation," the article ended with the description of three cases of "rejuvenation" in humans. Without their knowledge, three male patients (aged forty-four, sixty-six, and seventy-one) were subjected to ligation of their seminal vesicles. Reportedly, in a few weeks, the three individuals showed marked improvements in their health. The acute signs of old age—facial lines, crooked posture, and bodily weakness—began to disappear. The patients regained vitality, energy, appetite, muscle tone and strength, as well as sexual potency.

In the fall of the same year, Serge Voronoff (1920), a surgeon at the College de France in Paris, published a nearly three-hundred-page-long monograph under the enticing title *Vivre! Etudes des moyens de relever l'énergie vitale et de prolonger la vie*. The monograph detailed Voronoff's experiments with the transplantation of sex glands from young to old animals (sheep and goats). Illustrated by photographs of the animals before and after the operation, Voronoff's book claimed that the transplantation led to the "rejuvenation" of the old animals. The book also included a description of two cases of successful transplantation of the thyroid gland from apes to humans and concluded with an enthusiastic prophesy that the transplantation of sex glands from apes to humans would become a certain means of rejuvenation.

Steinach's and Voronoff's publications caused a furor in both academic and lay circles around the world. Among the many countries seized by the new "rejuvenation craze," none seemed to provide a less likely locale for the experiments with sex gonads than Russia. Torn to pieces by a bloody civil war, ridden with epidemics, and nearly starved to death, the country was on the brink of economic and political collapse. Yet its medical and scientific communities, as well as the general public, enthusiastically embraced the news of rejuvenation experiments.

Within a few months of their original publications, abstracts of both Steinach's and Voronoff's works appeared in Russian translation, accompanied by a barrage of articles in professional and popular periodicals. With the end of the civil war and the revival of the Russian publishing industry, the country witnessed a virtual explosion of publications on rejuvenation. Already in 1922, two publishers—one in Petrograd, another in Moscow—issued two different translations of a brochure, entitled *Rejuvenation and Prolongation of Individual Life,* written by Paul Kammerer (1921, 1922a, 1922b), Steinach's ardent supporter and colleague at the Vivarium.[8] In the subsequent three years, 1923-25, Russian biologists and physicians produced several dozen books and more than one hundred articles on rejuvenation.[9] Nikolai Kol'tsov (1923b, 1924b), director of the Narkomzdrav Institute of Experimental Biology, compiled two volumes of translations with key works by Steinach and Voronoff, as well as by other continental biologists and clinicians who had continued and expanded the research of the two pioneers. He also published detailed surveys of the newest American literature on the subject (Kol'tsov 1923a, 1924a). Boris Zavadovskii (1923), a professor of biology at the Sverdlov Communist University in Moscow, published a book-length survey entitled *The Problem of Senescence and Rejuvenation in Light of the Doctrine of Internal Secretions*. Five different translations of Voronoff's books came out in Moscow, Petrograd, and Kharkov (Voronov 1924a, 1924b, 1924c, 1924d, 1925). A well-known Petrograd publisher, Practical Medicine, also issued a translation of a voluminous monograph, *The Theory and Practice of Rejuvenation,* written by Peter Schmidt (1922, 1923), one of the most active German proponents of Steinach's operation.

As one might expect, the popularization of Steinach's and Voronoff's works stimulated attempts at replicating their operations by a number of Russian physicians and biologists. As early as 1922, records of such attempts appeared in Russian medical literature (Girgolav 1922). In the spring of 1924, another Petrograd publisher, Medicine, issued a special collection entitled *Rejuvenation in Russia.*[10] The hundred-fifty-page volume included six articles detailing the theoretical foundations for, laboratory research on, and clinical results of Steinach's and Voronoff's operations. According to the data presented in the volume, Russian surgeons had performed nearly two hundred operations by the time it went to press in late 1923. But, of course, a single volume could not even begin to represent the actual scale of the "rejuvenation craze." In fact, in Moscow and Petrograd, Kharkov and Smolensk, Baku and Kazan, Tashkent and Omsk, Tiflis and Odessa, surgeons performed Steinach's and Voronoff's operations on humans, while biologists experimented on a variety of animals, including cats, rabbits, dogs, and horses.[11] Practically all medical and biological periodicals, ranging from the *Russian Physiological Journal* to the *Archive of Clinical and Experimental Medicine, Advances in Experimental Biology,* and *Physicians' Affair* carried articles on rejuvenation.[12]

In 1923–25, meetings of various professional societies and periodic conferences of Russian biologists and physicians hotly debated rejuvenation issues. At the Sixteenth Congress of Russian Surgeons, in May 1924, Steinach's and Voronoff's operations, and the transplantation of endocrine glands more generally, became a major subject of discussion.[13] Reports on rejuvenation were delivered at the Second Scientific Congress of Central Asian Physicians in Tashkent, at regional conferences of Ukrainian surgeons, and at meetings of Georgian physicians.[14] The

Seventh All-Union Congress of Internists in May 1925 held a special plenary session on the transplantation of endocrine glands with twelve reports.[15] The Society of Russian Surgeons in Moscow and the Pirogov Society of Surgeons in Petrograd regularly listened to reports on—and observed patients who had undergone—"rejuvenation" by Steinach's and Voronoff's methods, as did the members of the popular forum of Petrograd physiologists, Physiological Conversations.

The debates over Steinach's and Voronoff's operations in Russia were as diverse as those elsewhere. Some physicians and scientists rejected Steinach's or Voronoff's operation (or both); others enthusiastically endorsed them.[16] Some reported complete success in achieving "rejuvenation" by Steinach's and/or Voronoff's technique. Others claimed only limited success, describing the effects produced as temporary and arguing that the operations resulted not in a "true rejuvenation" but only in a certain "refreshment" of vitality.[17] Still others announced complete failure, recording "no effect whatsoever."[18] But these disagreements only further fueled the public fascination with "rejuvenation."

The excitement and polemics about rejuvenation quickly spilled over from professional periodicals and conferences to the public scene. All available media became engaged in the popularization of rejuvenation. During 1923–25, a series of public disputes and lectures on the subject took place at Moscow's major venues for scientific propaganda: the Polytechnic Museum and the House of Scientists. In April 1923, Kol'tsov delivered a public lecture on "the newest works in the field of human surgical rejuvenation" at the House of Scientists. In February 1924, the Polytechnic Museum hosted a "public dispute" entitled "Is Rejuvenation Possible?" moderated by Zavadovskii.[19] In December 1924, the Polytechnic Museum held another "public dispute" on "rejuvenation" with Zavadovskii as a keynote speaker. In February 1925, Kol'tsov's lecture "Wondrous Achievements of Science" was broadcast all over the country from a radio station in Moscow. In his lecture, Kol'tsov (1925) singled out current research on rejuvenation as one of the most impressive achievements of modern science.

The printed and spoken word was supplemented with illustrations. In the fall of 1923, two German-made motion pictures about Steinach's research were shown in Moscow.[20] In June 1924, the House of Scientists held a special viewing of Steinach's film *Rejuvenation*, with extensive commentary delivered by Kol'tsov.[21] The same year, Narkomzdrav's widely publicized permanent exhibition on "the protection of health" in Moscow featured a special section on rejuvenation,[22] while a state publisher issued a set of thirty-seven slides on rejuvenation for "cultural propaganda" in workers' clubs, schools, and libraries.[23] Not to be outdone by their German colleagues, Russian proponents of rejuvenation also produced a motion picture about their research.[24] Entitled *Who Needs to Be Rejuvenated*, the film hit Moscow movie theaters in the fall of 1925, accompanied by public lectures on rejuvenation and press interviews by its producer, the well-known physiologist Leonid Voskresenskii.[25]

Practically all popular weeklies and monthlies carried the latest "news" on rejuvenation research.[26] An article on rejuvenation opened the first issue of a new magazine, *Science and Technology*, in February 1923 (Vasilevskii 1923). *Hygiene and Health of the Worker Family*, another new magazine established in September 1923, followed suit (Iul'ev 1923). Many other new weeklies that appeared in 1923–25, such as *Flash* and *Red Field*, regularly published articles on the subject (e.g., Pyzhov 1923). Even such literary journals as *Red Virgin Soil, Star, New World*, and *Young Guard* popularized rejuvenation (Zavadovskii 1921; Vasilevskii 1924; Kharin 1924; and Shmidt 1925).

The country's central newspapers, *Pravda* and *Izvestiia*, became engaged in the popularization of Steinach's and Voronoff's work, publishing articles by and interviews with the leading Russian "rejuvenators," Kol'tsov and Zavadovskii.[27] Other newspapers, ranging from the *Workers' Gazette* to the *Whistle*, followed suit.[28] In December 1923, the Moscow City Soviet established a new daily, *Evening Moscow*, which became one of the most popular newspapers in the capital and beyond. Unlike other government newspapers, such as *Pravda* or *Izvestiia*, the new daily did not publish decrees and political propaganda. It was a typical "yellow" paper that brought to the Moscow public a variety of juicy rumors, crime chronicles, and short (usually humorous) fiction, along with information on sports and cultural events. Among various subjects of popular interest, rejuvenation figured prominently on the pages of *Evening Moscow*. In 1924–25, the newspaper carried an item on the subject almost every month, reprinting "information" from foreign newspapers and publishing interviews with Russian proponents of rejuvenation.[29] A particular flavor of these publications can be illustrated through the following two examples. In September 1924, *Evening Moscow* reported that former British prime minister Lloyd George was planning to travel to Paris to undergo rejuvenation in Voronoff's clinic. A few days later, it informed readers that to date Voronoff had "rejuvenated" nearly fifteen hundred people.[30] In 1925, the newspaper carried fifteen items on rejuvenation, ranging from a notice on the death of Baron Rothschild, who had been "rejuvenated" just a few months earlier, to a report on recent rejuvenation operations performed by Moscow surgeons.[31] It also became a major source of "information" for many provincial newspapers, which

Эндокринный завтрак в Боткинской
больнице.

5.1 A cartoon on rejuvenation à la Voronoff. The caption reads "An
Endocrine breakfast at the Botkin Hospital." (From *Ne novyi
khirurgicheskii arkhiv* [*Not New Surgical Archive*, a humorous
supplement to the *New Surgical Archive*], 1927, 1.)

regularly reprinted articles that first appeared in *Evening Moscow*. By the
end of 1925, "Steinach" and "Voronoff" had become household names,
while "rejuvenation" had become a fixture of Russian popular culture
and began to figure regularly in fiction and theater.[32]

So, by the time Bogdanov began working on his *Struggle for Viability*,
issues of aging and rejuvenation had become arguably the "hottest"
topic in experimental biology and medicine in Russia. Thus, reframing
his work on blood exchanges into a "general theory" of senescence
certainly afforded Bogdanov a good strategy for "selling" his ideas to
the scientific and medical communities, as well as to the general public.
Furthermore, such reframing was probably also quite "sellable" to his
party patrons. Available evidence indicates that during 1924–25, the Bolshevik "old guard" was a major beneficiary of rejuvenation operations in
Moscow. Furthermore, in late 1925, Narkomzdrav moved to establish its
own primate breeding station; one of the station's major goals was to
become a domestic supplier of "monkey glands" for transplantations
à la Voronoff.

THE TECTOLOGY OF SENESCENCE

In accordance with his tectological approach, Bogdanov began his treatise with a claim that to truly advance studies of senescence, scientists need to "formulate the issue in its most generalized form." The notions of "old age" and "youth" were insufficiently universal and must be replaced by the notion of "viability," which Bogdanov characterized in the following manner:

> In the struggle against its natural environment, an organism neutralizes the environment's hostile influences and overcomes the environment's resistance by expending its accumulated energy. The larger the amount of such accumulated energy and the better such energy is organized (which depends on the particular structure of the organism), the more successfully and more perfectly the organism accomplishes this task. Taken together, these two components [energy and its organization represent the measure of the organism's "strength" in its life struggle—the measure of its viability. (Bogdanov 1927a, 7)

From this perspective, old age is generally characterized by somewhat lower viability than youth. But in certain respects, the old organism is equipped better than the young one for life's struggles. Bogdanov used his favorite example of acquired immunity to infectious diseases, asserting that "in this direction, the viability of the old organism is still increasing." Thus, in its "usual understanding," the task of "rejuvenation" appears to be formulated incorrectly, since the real task is "to raise and maintain the organism's viability in all directions at once." This, according to Bogdanov, is the "most general," and hence the "simplest," way to formulate the issues of senescence, which would illuminate the "path for investigations of and solutions to all the partial tasks that emerge within its framework." Thus, Bogdanov continued, his "general theory of viability" is applicable to the tasks facing the entire field of medicine and hygiene, as well as to the specific task of the "struggle against the old age."

Bogdanov analyzed the two basic components of viability: *quantitative,* defined by the amount of energy available to the organism, and *structural,* defined by the particular organization and coordination of various subsystems of the organism. Characterizing life as "a form of dynamic equilibrium," he discussed the exchange of energy between the organism and the environment, detailing processes of assimilation, dissimilation, self-regulation, and self-replication as the basic characteristics of life. He concluded that "the organism, the self-replicating machine of life, is *a system of the equilibrium of equilibrium systems*," within which "the self-regulation of parts is complemented and completed by their mutual regulation within the whole" (Bogdanov 1927a, 26; italics in the original). He restated "the minimum principle," a major "tectological" principle he had discovered in his analysis of the organization of complex systems, asserting that the ultimate limit of the organism's viability is defined by the viability of its weakest part. In essence, Bogdanov's theory treated senescence as "a decline of viability" due to the accumulation of "imbalances" in both its quantitative and its structural components.

Bogdanov employed these basic principles of his analysis of "the life machine" in a review of the three current concepts of senescence and its mechanisms: endocrinological (the decline in the production of sex hormones), cytological ("internal" intoxication by waste products), and bacteriological ("external" intoxication by the products of intestinal bacteria). As one might expect, he concluded that each of the three concepts in and of itself could not possibly explain the complexity of the aging processes in various organisms (from protozoa to plants to animals), and each presented merely a particular case for his more general "tectological" theory of aging (Bogdanov 1927a, 50).

Accordingly, after a lengthy survey of various facts reported in the literature, Bogdanov concluded that the ideas and techniques of "rejuvenation" advanced within the frameworks of the three concepts to stave off, impede, or reverse aging processes were inefficient, insufficient, and inadequate (they would always remain partial). He suggested that all of the proposed techniques of "rejuvenation"—from Brown-Sequard's injections of testicular extracts to Steinach's vasectomies, and from Metchnikoff's "yogurt drinking" to Voronoff's sex gland transplants—could be subsumed under, supplemented by, and eventually replaced by blood exchange, which was a simpler, safer, more effective, and—most important—more "universal" technique.

Bogdanov began with a theoretical substantiation of this view, which in its basic arguments expanded on Netti's explanation in *Red Star*: both conjugation in unicellular organisms and sexual copulation in multicellular organisms lead, as a general rule, to a definitive increase in the viability of their "progeny." He discussed certain examples of both processes, ranging from the "immortality" of infusoria resulting from their conjugation to the increase in the size of walnuts resulting from the hybridization of different varieties (borrowing the latter example from Timiriazev). Bogdanov concluded that such increase in the viability of the "progeny" was the result of mutual compensation of certain imbalances in each of the "unified" organisms. Since it was highly unlikely, he surmised, that in the course of their individual lives the two cells had accumulated the same imbalances (or the same "minimums"), their "unification" would balance out their differences. But he also noted that conjugation and copulation could lead to the decrease of viability due to the mutual reinforcement of misbalances and illustrated such a situation with several hypothetical examples.

If nature increases the viability of organisms by conjugation, Bogdanov theorized, what can stop humans from deploying the same method, only appropriately modified to suit the specific human "organization," to increase their own viability? He asserted that conjugation of both two infusoria and two sex cells in "higher organisms" is only possible because of the "physical intermixing of their semiliquid colloidal protoplasm." For multicellular organisms, the only tissues that have this consistency and thus could serve the same purpose are blood and lymph. Therefore, Bogdanov drew the obvious conclusion: an exchange of blood (and/or lymph) between two multicellular organisms must, as a general rule, lead to the increase of their viability.

He reiterated his favorite thesis that blood is a "universal" tissue that "unifies" all other organs, tissues, and cells in the organism, bringing in oxygen, nutrients, and other "vital substances," such as hormones, and taking out various products generated by all the cells, tissues, and organs, including hormones, carbon dioxide, and "toxic waste." Therefore, the chemical composition of blood at every moment in life reflects the activity of the organism as a whole and not merely that of its "blood-producing" organs: bone marrow, spleen, and lymph glands. This view led Bogdanov to two basic assumptions. On the one hand, he surmised, as the external environment for the life of cells, tissues, and organs, blood itself might actually become the source of misbalance in the organism, if organs responsible for the removal of "waste" (kidney, liver, etc.) fail to function properly. Such a situation would definitely require that the blood of such an organism be "purified" or "renewed" by some artificial means, and, of course, the best way to do this would be through a blood transfusion or blood exchange. On the other hand, Bogdanov inferred, "blood is the bearer of individual, racial, group-specific, and species-specific characteristics of any organism, the same way its germ-plasm is." All of these various characteristics had to be taken into account during blood transfusions/exchanges. In this light, he discussed blood groups (and blood incompatibility more generally), various forms of immunity (acquired, hereditary, and transferred), and the inheritance of acquired characteristics illustrated by a few examples taken from Timiriazev's and Kammerer's works.

Having set up his theoretical framework, Bogdanov briefly surveyed the history of blood transfusions, focusing mostly on the recent advances in the technique, particularly the use of potassium citrate to prevent the coagulation of blood. He provided an overview of the most common applications of blood transfusions in current medical practice, including blood loss, shock, blood diseases, poisoning, burns, and septicemias, discussing each of the applications from the vantage point of his concept of quantitative and structural viability. He also reviewed animal experiments on the so-called joint blood circulation (in which the blood systems of two animals had been joined by the anastomosis and which he had found in Keynes's book), as a certain kind of "blood exchange," noting that these experiments confirmed his basic ideas, but this method could not be applied to humans.

Bogdanov outlined the history of his personal interest in blood transfusion, illustrating it with a long excerpt from *Red Star* and brief references to *Tectology*. Then he presented "experimental evidence" to support his theory: blood exchanges performed by his group during 1924–25 between four aged and seven young individuals. According to Bogdanov, all four older participants undoubtedly benefited from the exchanges, while the blood exchange exerted "positive influence" on the health of five of the seven younger "companions." Obviously responding to the problem of "Soviet exhaustion," he particularly emphasized improvements in "nervous state" and "ability to work" among all participants in the exchanges.

There was, however, one peculiarity of Bogdanov's experiments, which he did not address explicitly in his description: all of the exchanges had been performed between individuals of the same sex. This peculiarity certainly reflected Bogdanov's understanding of exchanges as the "balancing out of imbalances"—in this particular case, the imbalances in the quantities of sex hormones. Although Bogdanov dismissed the "hormonal" theory of aging as partial and insufficient, he obviously still believed (along with the majority of 1920s "rejuvenators") that a leading cause of aging was the "weakening" of sex glands. Theoretically, then, the young had "too much" and the old "too little" of sex hormones, and an exchange would level out this particular imbalance. Not surprisingly, in one exchange Bogdanov recorded exactly that result: "sexual activity" in the older male companion had increased, and in the younger one decreased (Bogdanov 1927a, 136). But the dominant view of the time was that "male" and "female" hormones were completely different and antagonistic.[33] This perhaps explains why Bogdanov did not go beyond same-sex exchanges and did not "risk" exchanging blood between companions of the opposite sex, even though at one point he did give his own blood to a younger female "companion" to compensate for the amount she had lost giving blood to an older female. He might have thought that even if the quantity of hormones transmitted during blood exchanges was not large, an exchange between opposite sexes could lead to a "canceling out" of hormones and thus to a "diminishing" of sex characteristics in both companions, as had "happened" on Mars in *Red Star.*

In the end, Bogdanov admitted that "the small number of cases" and "certain technical deficiencies in the experiments" prevented him from reaching any "decisive" conclusions. "Our experiments do not yet present sufficient proof for our theoretical predictions," Bogdanov declared, "yet something we did prove—namely, that *there is something here worthy of investigation*" (Bogdanov 1927a, 137; italics in the original).

The last chapter of his treatise, entitled "Prospects," was even more optimistic: "The method of exchange or substitution of blood, if it is correct in essence, must have significance not only for the struggle against the age-induced decline of life. *All* of the various forms of insufficient viability would be susceptible to its influence" (Bogdanov 1927a, 137; italics in the original). The last statement clearly implied that blood exchanges could become a preferred method of treating "Soviet exhaustion," which plagued the country's ruling elite.

Bogdanov identified two major lines of future investigation into the effects of blood exchanges: the "transfer of immunity" and the "equalization of extremes," providing several hypothetical schemes of possible experiments. He suggested, for example, that blood exchanges between the old and the young could be studied in relation to particular diseases that predominantly affect only one of the two age groups. "In the young, cancer occurs only in extremely rare instances," he theorized; "hence, one has serious reasons to think that young blood could be the best means in the struggle against cancer." Tuberculosis provided a complementary example: since the disease "occurred mostly among the young," blood exchanges could transfer the immunity that the old had developed to the young. Furthermore, Bogdanov continued, "there are certain reasons—at this point purely theoretical ones, but they could be checked out in practice—to think that the blood conjugation method would be able not only to defend the organism against its natural or pathological decline but could also lead to the positive expansion of the organism's activities and resistances, which are usually called 'strengths,' 'abilities,' and so forth" (Bogdanov 1927a, 140). Here Bogdanov clearly suggested that blood exchanges would speed up the psychological and physiological perfection of human beings as they had perfected the inhabitants of his fictional Mars.

Bogdanov conceded that "nowadays, the idea of fighting for the organism's general viability by blood transfusions is floating in the air." But, he argued, aside from his own, other attempts to apply that idea in practice had been "somewhat strange" and had had "dubious theoretical foundations." He singled out the research of the French physician Helán Jaworski on the effects of transfusions of small amounts of blood from young to old individuals.[34] Since Jaworski's work was completely unknown in Russia (I did not encounter a single reference to Jaworski in any other biological or medical publications of the time), it seems most likely that Bogdanov learned of Jaworski's work from . . . *Evening Moscow*. In October 1925, the newspaper carried a short report on Jaworski's "rejuvenation vaccine," highlighting his sensational claim that "to transform a feeble old man into a flourishing youth, it is sufficient to infuse him with young healthy blood" ("Perelivanie krovi" 1925c). In all likelihood, Bogdanov saw the article.[35] Obviously struck by the discovery that someone had already been pursuing his basic idea, Bogdanov launched a search of the French literature and found a short note, "On the Question of Rejuvenation," that Jaworski (1925a) had published in *Revue Scientifique* in the fall of 1925. But he could not find Jaworski's magnum opus that appeared a few months later under the title *La régénération de l'organisme humain par les injections de sang* (Jaworski 1925b) or his 1918 theoretical treatise underpinning the experiments, *Pourquoi la mort? L'intériorisation,* which in January 1926 came out in its second edition (Jaworski 1926). It seems likely that Bogdanov's friends in Bolshevik diplomatic circles (either Krasin, who had been the Soviet ambassador to France, or one of his underlings) supplied him with whatever was available in French bookstores at the time, including several recent books on blood transfusions and a copy of Voronoff's 1924 book on monkey gland transplants. Alas, they could not find Jaworski's books, for, as Bogdanov noted, they "were sold out in France." To illustrate his assessment of Jaworski's work, Bogdanov included in the text of his book a translation of Jaworski's 1925 article in its entirety. But despite glaring similarities between his own and Jaworski's theoretical arguments (to give but one example, Jaworski also referred to conjugation and the renewal of nutrient medium as sources of "immortality" in infusoria), Bogdanov found Jaworski's thoughts "strange and unconvincing, some kind of a speculative 'system.'" He reiterated that the general idea of rejuvenation by blood transfusions was correct and he had published it "ten years before Jaworski," but he rejected completely both the practical methods and the theoretical foundations of Jaworski's work.

Having dealt the "mortal blow" to a possible (though completely unknown to his audience) contender to his priority in discovering blood transfusions as the method of rejuvenation, Bogdanov then turned to the only published response to his "tectological" theory of senescence, which had appeared prior to his work on *Struggle for Viability*. As we have seen, having articulated his basic idea in *Red Star*, Bogdanov expanded it in the first volume of *Tectology* and further elaborated it in the "outlines" of *Tectology*'s final volume that had appeared in *Proletarian Culture*. But the ideas of blood exchanges "buried" in these publications went virtually unnoticed by Russian biologists and physicians. The majority of older, established Russian scientists and doctors had been hostile to the Bolshevik Revolution, had no interest in "proletarian culture," and most likely had never even looked at the movement's oracle. Since Bogdanov had never published in their professional periodicals, they remained completely oblivious to his exploits.

But the younger generation of biologists and physicians saw the Bolshevik Revolution as a great opportunity to advance their intellectual interests, institutional positions, and career ambitions. They actively demonstrated their loyalty to the regime by joining various "communist" societies and actively exercising the new "Marxist" lexicon in their own specialties. One such opportunist was Boris Zavadovskii. Born in 1895, he graduated from Moscow University in late 1917, just as the Bolsheviks seized power in Russia. Two years later, he became a professor of biology at the newly established Sverdlov Communist University. Under the university's auspices, he organized a biological museum named after Timiriazev and a large biological laboratory staffed with his "communist" students. In 1921–22, Zavadovskii published a series of literature surveys on recent developments in the study of senescence in various scientific journals and popular magazines. Trained as a physiologist, he was deeply interested in the endocrinology of growth and aging and lectured on both subjects at the university as well as at various public forums. In early 1923, Zavadovskii converted his literature surveys and lectures into a book-length treatise, entitled *The Problem of Old Age and Rejuvenation in Light of the Doctrine of Internal Secretions*. Perhaps Zavadovskii himself stumbled upon Bogdanov's publications, or, more likely, one of his "proletarian" students aware of the professor's interest in the subject pointed it out to him. Whatever the case, at the very end of his overview, Zavadovskii inserted a paragraph that briefly discussed Bogdanov's "outwardly scientific but in essence unconvincing attempt" at solving "the problem of old age," which had appeared in *Proletarian Culture* in 1920 (Zavadovskii 1923,119–20).

Zavadovskii noted that Bogdanov's "attempt" suffered from three basic shortcomings. First, it was based "on a number of false and incorrect—from the viewpoint of modern science—facts," such as "blended heredity" and "archaic notions of the inheritance of acquired characteristics." Second, it depended "upon extremely risky parallels and analogies," such as the analogy "between blood transfusion and conjugation or even a telepathic exchange of thoughts." Finally, "without any substantiation," it considered certain "hypotheses as firmly established facts"—for instance, that transfused "blood preserves its individuality in a different organism." These shortcomings, in Zavadovskii's opinion, gave Bogdanov's conception "a clearly speculative and dogmatic character and thus deprived it of any scientific value." He concluded by refusing to even broach a discussion of "how, in general, such deductive-metaphysical constructions could advance solutions to scientific problems."

Bogdanov was clearly infuriated by Zavadovskii's unequivocal dismissal of his major idea. In *Struggle for Viability*, he reproduced Zavadovskii's paragraph and launched his own seven-page-long (!) counterattack on Zavadovskii's assessment. Writing in a style typical of the Bolshevik intraparty polemics (the same style Lenin had employed in *Materialism and Empiriocriticism* and Bogdanov, in turn, had used in his response to Lenin's critique),[36] Bogdanov called Zavadovsky "an ignoramus," exclaiming dramatically: "a biologist is lecturing a physician, while having no idea about what has been done on the question of blood transfusions in medicine." Bogdanov refuted Zavadovskii's arguments point by point, focusing in particular on the inheritance of acquired characteristics. Although the issue was far from central to Bogdanov's own "tectological" theory, his choice of this particular focus was quite deliberate.

As I have detailed elsewhere, during 1923–26, the polemic over Lamarckian inheritance flared up in Soviet experimental biology and medicine (Krementsov 2010). On the one hand, it was a centerpiece in the struggles to legitimize the nascent discipline of genetics. On the other hand, it became a central issue in numerous efforts to "synthesize" Darwinism and Marxism undertaken under the umbrella of the Communist Academy and various "Marxist" societies, such as the "Society of Materialist-Biologists," "the Circle of Materialist-Physicians" and "Leninism in Medicine." In early 1926, the Communist Academy Presidium discussed plans to create a special laboratory to prove the validity of the inheritance of acquired characteristics and even invited

89

Paul Kammerer, notorious for his advocacy of Lamarckian inheritance, to head it. Kammerer accepted the invitation, but on the eve of his departure for Russia, faced with accusations of scientific fraud, he committed suicide, and the plan fell through.[37] Bogdanov, as a member of the Presidium, had clearly been aware of this plan, as well as of continuous debates in the academy on the inheritance of acquired characteristics.

Indeed, starting in the fall of 1925, the Communist Academy section of natural sciences held regular meetings to discuss the interrelations among Darwinism, Marxism, Lamarckism, and genetics. At one such meeting in late November 1925, the audience was treated to a two-hour report that advocated a "synthesis" of genetics and Lamarckism. And delivering the report was none other than Zavadovskii! Even if Bogdanov had not attended the meeting, Zavadovskii's report soon appeared in the academy's major periodical, *Under the Banner of Marxism,* providing Bogdanov with a perfect tool to undermine his authority as an impartial, objective critic, by pointing out that, contrary to his 1923 critique, Zavadovskii (1925) himself now supported the "archaic notion" of the inheritance of acquired characteristics.

Having taken to task the imaginary contender Jaworski and his only critic Zavadovskii, Bogdanov turned his attention to the real competitors: Russian transfusiologists. Obviously, in order to justify his claims to priority in introducing the modern techniques of blood transfusion to Soviet medicine, Bogdanov appended the book with a five-page critical review of two recent monographs on the subject, one written by Nikolai Elanskii and another by Iakov Bruskin. Bogdanov conceded that Elanskii's monograph was "useful," though "superficial and occasionally mistaken." He saved his most vehement criticism for Bruskin's monograph. Bogdanov ridiculed Bruskin's preference for direct transfusions, stating that "in the countries where blood transfusions have been most fully developed, this technique is disappearing" (Bogdanov 1927a, 155–56). He dismissed outright Bruskin's critique of transfusions with citrated blood and his experiments on dogs conducted to demonstrate the toxicity of potassium citrate: "in their faulty design, Dr. Bruskin's cruel experiments are completely useless for science, and in his mistaken interpretation, they can only do harm."[38]

Bogdanov finished his book on a high note:

> In our epoch, individualist culture dominates; its atmosphere is unfavorable to both our method and our viewpoint that underpin the method. Their true foundation, labor collectivism, is only beginning to come to life. When it triumphs, then those difficulties and obstacles that now stand in the way of *physiological collectivism* will be removed, [and] then this

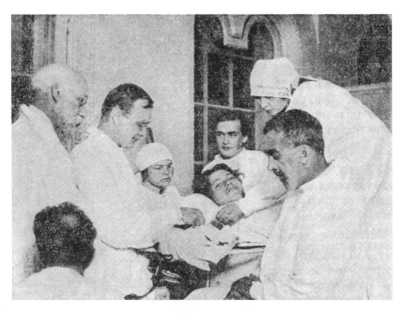

5.2   A blood transfusion at Bogdanov's institute, circa 1928: Semen Maloletkov
(*standing on the left*), Dmitrii Gudim-Levkovich (*sitting on the right*). (From
30 dnei [30 Days] 7 [1928]: 76.)

collectivism will blossom. As for now, it is not enough to wait for this
victory; one needs to exercise strong will for it [to come]. Difficulties and
obstacles exist to be overcome. To work, to investigate! (Bogdanov 1927a,
154; italics in the original)

A few weeks after the release of *Struggle for Viability*, Bogdanov
(1927c) published yet another article in *Izvestiia* under the expressive
title "The Small Beginning of a Large Affair." The article was a thinly
veiled advertisement for his manifesto. It simply recounted once again
the "tasks and aims" of his institute outlined in his article published in
the same newspaper the previous year, together with Bogdanov's ideas
about blood exchanges elaborated in his book.
   Bogdanov's passionate call to arms in the field of blood transfusion
was not left unanswered.

6 Blood and Socialism

THE DEATH OF A HERO

Bogdanov's vision of "physiological collectivism" forcefully articulated in *Struggle for Viability* seemed perfectly suited to the atmosphere of revolutionary enthusiasm that permeated 1920s Russia. Furthermore, Bogdanov had at his disposal a large research establishment to bring his vision to life through experiments with blood exchanges. In July 1927, an article on the "great social-practical and research tasks" of Bogdanov's institute in the Central Committee's mouthpiece *Pravda* forecast: "Here begins what in the future will create physiological collectivism, which will allow the human organism to rely not only on its own forces and means in its struggle against diseases, wounds, and a general deterioration of life but also on the elements of viability developed by other human organisms" (Baranovskii 1927). Yet, as Bogdanov had himself foreseen, for this prediction to come true, he needed to overcome many "difficulties and obstacles." He singled out "individualist culture" as the most difficult barrier in the way of his vision. Life proved him wrong: the major obstacles lay elsewhere.

DETRACTORS AND SUPPORTERS

Bogdanov did not have to wait long for public reaction to his manifesto. At the end of December 1926, *Izvestiia* carried a lengthy review of Bogdanov's book written by a staff reporter (Ottenberg 1926). Given Bogdanov's long association with the newspaper's editor in chief Skvortsov-Stepanov, the review was probably prearranged and its general congratulatory tone predetermined. "The strictly scientific elaboration of the topic, simple and clear delivery [of material], new and original approach to the question of the struggle against old age—these are the basic qualities of the book," raved the reviewer. Summarizing Bogdanov's theory and his "not numerous, but interesting in their results, experiments" on blood exchanges between the young and the old, the reviewer called upon "skeptical scientists" to take part in the development of issues raised by Bogdanov's work. "We recommend this book to be read not only by every physician and biologist but also by all readers who have some knowledge of natural sciences," the reviewer concluded.

It is impossible to know how many people with "some knowledge of natural sciences" actually read the book. But some physicians certainly did, and the reviews that appeared both in newspapers and in specialized medical periodicals were far from laudatory. The first critical review appeared in January 1927 in . . . *Evening Moscow*! The reviewer, identified only by the initials "N.N.," briefly recounted the book's contents and stressed that its author "had insufficiently elaborated the medical-practical side of the issue" (N.N. 1927).[1] One of the most widely circulated medical journals, *Physician's Gazette,* published a brief review by Mikhail Breitman (1927), a well-respected professor of internal medicine at the Leningrad Institute for Doctors' Continuing Education and a leading specialist in the diseases of endocrine glands. Breitman simply repeated word for word Bogdanov's preface to the book. Then, without any comment, he reproduced the book's table of contents, which—peppered with such phrases as "definite and indefinite imbalances of life coordinations"—probably appeared nearly incomprehensible to his audience. He pointed out Bogdanov's "sharp critique" of Bruskin's work and noted that the book contained quite a few "polemical attacks" on other researchers and often bore "the unnecessary character of a feuilleton." As regards Bogdanov's original experiments on blood exchanges, Breitman simply reiterated Bogdanov's own statement about "their small number and the scientific-technical deficiencies in their conception and performance," as well as Bogdanov's expressed belief that "physiological collectivism" deserves further investigation.

The *Kazan Medical Journal* published an even shorter review by N. Sokolov (1927), one of the lecturers at the Kazan University Medical School. Sokolov simply restated Bogdanov's general idea that the three current concepts of senescence "are partial and one-sided" and that blood exchanges should become the preferred method of the struggle against the "weakening of the organism." Judging from the records of Bogdanov's "first (10) experiments with mutual blood transfusions," the reviewer noted, "some [of them] could be considered successful." Sokolov also pointed out that "along with certain firmly established facts, much of the book presents 'risky creativity.'" Although the book "is easy to read," the reviewer concluded, "the unrestrained tone of its author's polemics with Zavadovskii and his harsh critique of Bruskin's work create an unpleasant dissonance."

Another leading journal, *Russian Clinics,* took a somewhat different stance. In its pages, Ivan Rufanov, a rising Moscow surgeon, published side-by-side reviews of the three recent books that dealt with blood transfusions: Elanskii's, Bruskin's, and Bogdanov's (Rufanov 1927a, 1927b, 1927c). He characterized Elanskii's and Bruskin's monographs as timely, informative, and valuable. He especially praised Elanskii's book, joining Fedorov's opinion that Elanskii "speaks of blood transfusions soberly, objectively, and dispassionately and gives the practicing physician all the opportunities to deal with this method of treatment accordingly." Even though he questioned Bruskin's unequivocal rejection of the citrate method and noted some typographical errors and misspellings of foreign names in his book, Rufanov concluded that "everyone interested in blood transfusions will find a lot to learn from Bruskin's book." Rufanov's take on Bogdanov's book was quite different. He avoided the evaluative statements that speckled his analyses of Bruskin's and Elanskii's monographs, limiting his review to a brief summary of the book's contents.

Nikolai Elanskii (1927) was much less reserved. He published a scathing review of Bogdanov's book in the *Journal for Doctors' Continuing Education.* He began with a simple statement that Bogdanov treated "the issues of blood transfusion from the viewpoint of general biology" while ignoring "the proper medical side of the issues." Elanskii criticized both the theoretical foundations of and the actual experiments on blood exchanges. He dismissed Bogdanov's key theoretical analogy between the conjugation of infusoria and blood exchanges among humans, since "infusoria and humans stand on opposite sides of the biological scale." But it was Bogdanov's actual experiments that elicited Elanskii's most extended critique. Noting the absence of preliminary animal trials and the very small number of actual blood exchanges, Elanskii assessed Bogdanov's experimental design step by step. He dismissed as unsubstantiated Bogdanov's insistence on indirect transfusion with citrated blood and his rejection of methods for direct transfusion, stating that "modern apparatuses for direct transfusion make this method simple in technique and absolutely safe for both donor and recipient." He emphasized that in assessing the results of blood exchanges, Bogdanov had relied on "general impressions," the "subjective feelings of the participants," and their "nervous state," while "the protocols of his experiments contain no mention of measuring temperature and blood pressure, urine analysis, blood cell count, analysis of respiration, blood chemistry, and so forth."

Elanskii was especially dismayed by Bogdanov's assertion that "young companions" had also benefited from blood exchanges with aged and often sick individuals: "If one looks at the experiments without bias, one cannot exclude the possibility that instead of imaginary immunity, the healthier companion receives real infection." He pointed out several instances of deteriorating health status of the younger companions recorded in Bogdanov's protocols, which might well have been manifestations of such secondary infections, but which Bogdanov himself had dismissed as "accidental" phenomena. "Like [Bogdanov's] theory, so too the experiments recounted to support it raise very serious objections," Elanskii insisted. He actually likened Bogdanov's experiments to the famous anecdote about the blood transfusion performed on Pope Innocent VIII in 1492. In Elanskii's opinion, Bogdanov's book "brings us back to ancient and medieval times, when a particular miraculous force, which helps in incurable illnesses and rejuvenates the elderly, was ascribed to blood in the same manner as Bogdanov does nowadays." "Blood transfusions have begun to attain scientific importance only after the exact medical indications and conditions [for this operation] have been established," Elanskii concluded, implying that Bogdanov's work clearly fell outside the exactitude of modern medicine.

If the medical community did take notice (however critical) of Bogdanov's incursion into their domain, the biological community seems to have simply ignored it. I did not find a single review of his book in any of the country's numerous biological journals.

Apparently, neither Bogdanov himself, nor his patrons in Narkomzdrav and the Central Committee, paid much attention to the opinions of leading Russian physicians and transfusiologists. In the late fall of 1927, Bogdanov produced the first public report on the actual work of his institute. Entitled *The First Year of Work by the Institute of Blood Transfusion, 1926–1927*, the forty-page brochure repeated again Bogdanov's favorite claims that prior to the establishment of his institute, blood transfusions had been virtually unknown to Russian doctors, that the operation "had been practiced by only a very few surgeons in the large centers, quite rarely, practically without any descriptions in the [medical] literature, and [it was] completely neglected even in the best clinics and largest hospitals."[2] Bogdanov announced that to date the institute staff had performed "more than two hundred transfusions." But he did not provide any serious (not to mention statistical) analysis of this work. He did not even group the transfusions according to patients' age, sex, occupation, or medical conditions that warranted the intervention, such as anemia, "Soviet exhaustion," or poisoning. He did, however, divide his account into two sections: "the results of transfusions along the lines pointed out by Western science" and "the results of transfusions along the lines of the institute's own research." But he failed to mention how many transfusions (out of the "more than two hundred" conducted) fell within each line of investigation. In the first section, Bogdanov briefly reported eighteen case histories of patients (whose ages ranged from two to sixty-two years) with a variety of conditions (from TB and anemia to burns and poisoning), giving anecdotal evidence on the effects of the blood transfusions. In the second section, which occupied nearly half of the entire brochure, Bogdanov described in a similar manner the results of nine blood exchanges.

The last three pages of the report dealt with "popularization" and "the institute's expansion and prospects." Noting that "propaganda of blood transfusions is a major task of the institute," Bogdanov provided a list of publications and reports produced by the institute personnel. The list shows that Bogdanov and his staff had made no attempt to address the institute's intended audience: practicing doctors. All the publications listed had appeared in daily newspapers and popular magazines, with not a single article in professional medical periodicals. Bogdanov had delivered a report on the importance of blood transfusions for the Red Army to a staff meeting of the Society for the Assistance to the Aviation and Chemical Industries in the USSR, but neither he himself nor anyone else on the institute's staff had taken any part in the numerous professional meetings held by Russian physicians during 1926–27, whether congresses of surgeons, internists, and gynecologists held at the time, or the numerous meetings of medical societies. Even though the institute had provided standard sera for blood typing—upon request and free of charge—to those doctors who came to visit, the report indicated that the courses on blood transfusions the institute had intended to hold for physicians never materialized. Similarly, the production of the "simplest apparatus" or a "comprehensive manual" for blood transfusions had not even been contemplated.

But, according to the report, the institute did grow quite considerably. In the summer of 1927, the institute created "a laboratory for animal experiments and biochemical analyses" headed by Aleksandr Bogomolets (1881–1946), a well-known pathologist and physiologist (and a longtime Bolshevik sympathizer),[3] who also replaced Maloletkov as the "deputy director for scientific affairs." Bogomolets had published widely on the issues of aging and endocrinology and was the first staff member with extensive experience in laboratory research. He brought to the institute two assistants, who immediately launched a program of investigations on dogs and rabbits regarding the role of blood transfusions in maintaining the organism's hormonal balance.

The institute also established an "experimental clinic" with five beds to study various blood diseases, particularly in relation to occupational

97

hazards. Kharlampii Vlados (1891–1953), a rising-star hematologist from Moscow University who had just coauthored the country's first (four-hundred-page-long) manual on "clinical hematology," was invited to head the clinic.[4] The institute acquired its own X-ray department with both diagnostic and treatment equipment, and it hired specialists in venereology, ophthalmology, and roentgenology, along with several additional surgeons and internists. The institute's operational budget for the next fiscal year more than doubled, reaching the staggering amount of 160,000 rubles.

What Bogdanov's glowing report did not mention was a serious mishap that had occurred in the institute's clinical work.[5] In late 1926, after two blood transfusions, one of the institute's patients showed symptoms of a syphilitic infection. An internal investigation uncovered that neither the two donors nor the recipient had been tested for syphilis prior to the transfusions. The donors had only been "checked by an experienced doctor, though not a specialist in venereal diseases." It was this mishap that prompted Narkomzdrav to allow the institute to hire a specialist venereologist and to expand its clinical laboratory (which had been responsible for conducting the Wassermann test for syphilis on donors).

Bogdanov also did not mention in his public report that he himself had an ongoing row with his deputies "for administrative affairs." The first deputy appointed by Narkomzdrav in May 1926, D. L. Zeilidzon, was an old member of the Bolshevik Party from Baku. He immediately tried to fill the institute with his acquaintances from Azerbaijan, and when Bogdanov refused to hire them, he started to spread rumors about the "political face" of the institute's personnel, implying that Bogdanov and his friends were all "anti-Bolsheviks." Bogdanov managed to convince Semashko to fire him. But the next appointee—Evgenii Ramonov, another old party hand with a medical background—proved to be an even worse choice. Ramonov had also begun his party career in the Caucasus, and in the first postrevolutionary years, he proved instrumental in setting up medical services and institutions in Ossetia. In the early 1920s, he came to Moscow and became a party watchdog in Narkomzdrav. He actually "supervised" Lenin's surgical operation in April 1922 and, together with Semashko, cosigned daily bulletins on the Bolshevik leader's health. Appointed a deputy director, Ramonov obviously decided to take over not only the administrative but also the scientific and clinical affairs of the institute. According to Bogdanov, he ignored the director, interfered with the medical orders of the institute's physicians, and embezzled the institute's funds. Bogdanov appealed to Semashko for help several times, but even the commissar could not fire the party appointee: the decision had been relegated to the Narkomzdrav party cell, which only reaffirmed Ramonov's employment. In January 1928, Bogdanov used his last resort: in a letter addressed to Semashko, Stalin, and Bukharin, he threatened to resign his directorship. The threat proved effective: Ramonov was fired. Andrei Bagdasarov (1897–1961), a young physician who had graduated in 1923 from the Medical School of the Second Moscow University and since then had worked in the Narkomzdrav apparatus, was appointed to the post.

In early March 1928, Bogdanov delivered a formal report on the institute's work to the Narkomzdrav Administrative-Financial Directorate, enclosing his 1927 brochure as a supplement. On March 14, he addressed a special meeting of the directorate's officials gathered to discuss the accomplishments and future plans of his institute.[6] The participants approved Bogdanov's report. In a special resolution, they emphasized that the "organizational period" in the institute's life was over and that the institute was ready "to expand and deepen its research agenda." The resolution recommended that the institute be transferred to the Narkomzdrav Clinical Division and renamed, according to Bogdanov's proposal, "the State Research Institute of Hematology." The resolution also called on Bogdanov to "integrate the work of the institute's experimental and clinical sections," to "expand the institute's contacts with other medical institutions," and to "prepare a syllabus for a special course on blood transfusions for doctors' continuing education." Although he squabbled over the wording of certain phrases in the resolution, Bogdanov was apparently quite satisfied with the outcome of his first official report to his patron: he was busily preparing the first volume of the institute's *Proceedings*, scheduled for publication in May.

Ten days after his report, on March 24, Bogdanov took part in yet another, his twelfth, blood exchange with a twenty-one-year-old male student. A few days earlier, a group of eight students from Moscow University had come to the institute and volunteered to take part in experiments with blood exchanges. They had undergone a medical checkup, but for one reason or another, they all had been turned away. Yet one student was called back to the institute for a meeting with its director. Bogdanov proposed that he and the student take part in a mutual blood exchange. The student suffered from an inactive form of tuberculosis, and Bogdanov had always believed himself to be immune to the disease. By exchanging blood with the student, Bogdanov obviously aimed at proving his idea that "younger companions" do benefit from blood exchanges with older ones due to the transfer of immunity the latter had inherited or had acquired during their lifetime. The new exchange also fitted perfectly with the plan of investigations Bogdanov had outlined in *Struggle for Viability*, which specifically singled out tuberculosis as a

6.1a

preferred research subject. It is also possible that Bogdanov had read Elanskii's review of his book and seized an opportunity to prove his critique wrong. But, despite the fact that the student had the same blood group as Bogdanov's, three hours after the exchange of nearly one liter of blood, both developed an acute adverse reaction. The student recovered. Bogdanov died two weeks later from what his doctors described as extensive "hemolysis" that led to renal and liver failure.[7]

THE DEATH OF A HERO

Bogdanov's death was highly publicized in the press as the last heroic act of an unselfish physician and revolutionary.[8] He was accorded a

6.1a–b Bogdanov's funeral, April 1928. (Courtesy of ARAN.)

state funeral.[9] Members of the country's highest elite paid their respects to the fallen comrade: Commissar Semashko and Politburo
member Bukharin delivered passionate speeches at Bogdanov's open
grave. Both eulogies appeared in print immediately in both *Pravda* and
Izvestiia, while Commissar Lunacharskii published an eloquent obituary
in *Pravda* (Semashko 1928a, 1928b; Bukharin 1928; Lunacharski 1928).
A week after the funeral, his institute was renamed the Bogdanov State
Scientific Institute of Blood Transfusion—one of the highest accolades
reserved for the ultimate Soviet heroes—and his widow received a state
pension for life.[10] In early 1929, the country's most popular adventure
and science-fiction magazine, *Around the World,* reissued—as a supplement to its regular subscription—both *Red Star* and *Engineer Menni*
with a print run of 120,000 copies each, far larger than any previous
printings![11]

Media hype surrounding his death and massive reprinting of his
novels notwithstanding, Bogdanov's death delivered a fatal blow both to
his vision of "physiological collectivism" and to his research program on
blood exchanges. Even though the Institute of Blood Transfusion now
bore his name, blood exchanges quickly faded both from the institute's
research agenda and from publications on blood transfusions in Soviet
popular and specialized periodicals. The writings of one particular ad-
mirer of Bogdanov's works, physician and well-known popularizer Lev
Vasilevskii, who himself had taken part in a blood exchange, illustrate

this process perfectly.[12] A few weeks before Bogdanov's death, Vasilevskii (1928a) published a laudatory article about his institute in the *Leningrad Medical Journal,* praising the exchange method as a "principally new" direction in blood transfusions, one that "had already proven itself in the clinic." In Bogdanov's obituary, expressively entitled "Death at a Scientific Post," Vasilevskii (1928d) again lauded "physiological collectivism" and portrayed the last blood exchange that had led to Bogdanov's death as "one—modest in appearance, but great in idea—of the countless cases of heroism in the history of medicine." In a survey article on blood transfusions published in the same issue of the journal immediately after the obituary, however, Vasilevskii (1928b) did not even mention Bogdanov's name or his theory.[13]

In late May 1928, as scheduled, the first volume of the institute's *Proceedings* came out under the evocative title *On a New Field* (Bogdanov, Bogomolets, and Konchalovskii 1928). Opening with Semashko's and Bukharin's eulogies, the volume carried a detailed clinical analysis of Bogdanov's illness and death following the fateful experiment, complete with the results of an autopsy, written by Maksim Konchalovskii, a prominent Moscow internist who had joined the institute's staff a few months earlier. At the end of his report, Konchalovskii posed the question directly: "Did Bogdanov's death crush the idea, which had given the foundation to all his life, which had provided focus to all the efforts of his creative philosophical mind, and which had so seduced and captivated his warm and sensitive heart?" Although Konchalovskii did not give a straight answer to his own question, he admitted that "the properties of blood of each individual cannot be represented by identical formulas. These properties are deeply different and individual." Nevertheless, he finished his report on a high note: "The idea of exchange transfusions cannot be abandoned. We, the workers of the institute that bears the name of its deceased director A. A. Bogdanov, should work tirelessly and spare no effort to ensure that his idea receives thorough scientific elucidation and gives humanity those benefits of which A. A. Bogdanov dreamed" (Konchalovskii 1928, XXV).

But even Bogdanov's successor as the institute's director, Bogomolets, quickly "forgot" the dreams of its founder. In a brief commemorative article entitled "On the Question of the Scientific and Practical Importance of the Technique of Blood Transfusion," which followed Konchalovskii's postmortem analysis in the institute's *Proceedings,* Bogomolets (1928) spared no words in praising Bogdanov's ideas and experiments. But a plan for the institute's work he submitted to Narkomzdrav in February 1929 did not even mention blood exchanges.[14] On March 21, 1929, *Evening Moscow* carried an article commemorating the anniversary of Bogdanov's death. Signed "Physician" (and perhaps written by Vasilevskii), the article again posed the question: "Did Bogdanov's death undermine the idea of blood exchanges?" (Vrach 1929). The answer given by the "Physician" was unequivocal: "in no way." "The ever-growing activity of the institute—as under Bogdanov, so too after his death—shows the great viability and fertility of his idea," the "Physician" insisted. On the same page, however, the newspaper carried another article written by Bogomolets (1929b) on research conducted at the institute, which made no mention of Bogdanov or blood exchanges.

Furthermore, just a few months later, Bogomolets (1929a) published a much longer article, under practically the same title he had used in the first volume of the institute's *Proceedings,* "On the Scientific and Practical Importance of the Technique of Blood Transfusion," but completely different in tone and content.[15] Bogomolets paid Bogdanov lip service by proclaiming that "he had managed to guess the outlines of scientific truth and—through his work at the Institute of Blood Transfusion—to open a new page in the doctrine of the scientific and practical importance of transfusions." But, Bogomolets stated bluntly, Bogdanov "had simplified to the extreme" the complex physiological and pathological processes in the human organism, and his views "did not correspond to modern medicine." Characterizing Bogdanov's ideas as "mechanistic," "schematic," and at best "hypothetical," Bogomolets portrayed Bogdanov's concept of "mutual increase in viability" through blood exchanges between the old and the young as "medieval mysticism mixed with the dogma of mechanistic materialism." He insisted that "details of Bogdanov's conception of the mechanism of the beneficial influence exerted by transfused blood in cases of the organism's premature wearing off had no proof in experimental research."

In early 1930, the Bogdanov Institute finally issued a long-promised manual for blood transfusions produced by the institute staff. Entitled *Blood Transfusion as a Method of Medical Treatment,* the collection addressed neither Bogdanov's ideas about the possible increase of "viability" nor blood exchanges (Bogomolets, Konchalovskii, and Spasokukotskii 1930).[16] By the end of the year, all that remained of Bogdanov's legacy in blood transfusion research and practice was his name attached to the institute he had built in Moscow.[17]

Bogdanov's death did not undermine the idea of blood transfusions per se, though it obviously did raise the issue of their safety in the public's mind. Two days after Bogdanov's death, *Evening Moscow* published an interview with Bogomolets under the telling title: "Is Blood Transfusion a Dangerous Procedure?" The first question the reporter asked Bogomolets was: "Could Bogdanov's death lead to a pessimistic assessment of blood transfusions in general?" Bogomolets, of course, answered the question in the negative, and the newspaper printed his words in bold type to reassure its readers that blood transfusions "*do not present any danger to those treated with this method.*"[18]

By the time of Bogdanov's death, the methods of blood typing, direct and indirect transfusions, as well as special apparatuses and accessories for the operation had been described in detail in numerous professional publications and had become familiar to many physicians. Some of them had mastered the techniques of blood transfusion and used transfusions from time to time in their daily practice. In early 1928, the Odessa transfusiologist Leon Barinshtein defended the country's first doctoral dissertation on blood transfusions, immediately issued as a voluminous monograph. Many enthusiasts envisioned blood transfusion as a standard clinical procedure readily available in any corner of the country. In implementing this vision, in August 1928, Commissar Semashko approved special guidelines for performing the procedure throughout the country. Published in Narkomzdrav's official journal, the guidelines clearly identified both the clinical indications and the technical requirements for the operation, though they did not include any mention of blood exchanges ("Instruktsiia . . ." 1928).

But, as some surgeons readily admitted, blood transfusion still remained a somewhat exotic procedure for the majority of practitioners.[19] Although Bogdanov had blamed the "predominance of individualist psychology" and the "ignorance of Western literature and experiences" for the slow diffusion of the technique among Russian clinicians, a stumbling block to the regular application of blood transfusions had been entirely different and was unmistakably recognized by Russian transfusiologists: the shortage of blood.

BLOOD AND SOCIALISM

Indeed, even before Narkomzdrav established Bogdanov's institute in Moscow, one of the early enthusiasts of the procedure, Leningrad surgeon Erik Gesse, had identified the problem. On October 21, 1925, at a meeting of the Pirogov Society of Russian Surgeons in Leningrad, Gesse demonstrated a patient who had been operated upon in connection with a bleeding stomach ulcer.[20] Gesse told the meeting that during the operation, he had transfused the patient with blood donated by his sons. The patient had fully recovered after the operation, and Gesse largely attributed this success to the transfusion. He lamented the absence of regular cadres of donors who could be called upon whenever the need for a blood transfusion emerged in the clinic. Gesse relayed to the audience that he "had lost one patient because I could not find a donor, and had lost a second patient because I could transfuse blood from only one donor, while the second donor had refused [to give blood] at the last moment, and the 300 milliliters of blood transfused [from the first donor] had turned out to be not enough [to ensure the patient's survival]." Gesse asked the meeting to endorse his idea of instituting a system of "donorship" (*donorstvo*) similar to "the one existing in America." "Because of the extremely unsatisfactory situation with donors," Gesse urged the society "to convene a special meeting" on the subject together with legal experts, representatives of the City Department of Health Protection, and representatives of Narkomzdrav. Gesse's idea found support among Leningrad surgeons: the society agreed unanimously that the "question of donorship is urgent and extremely important."

While Bogdanov was beginning renovations of the Igumnov mansion, in early May 1926, the Leningrad Regional Department of Health Protection held a special joint meeting of Leningrad physicians and department officials to discuss Gesse's report on "the organization of professional donorship in relation to the operation of blood transfusion" (Gesse 1926a). Gesse opened his address by stating that "the operation of blood transfusion has come out of the experimental stage and should be recognized as a scientifically sound and extremely important clinical procedure, which no appropriately functioning medical institution can do without." He referred the audience to his own and his assistants' publications on the subject, as well as to works by Shamov and Elanskii, for elucidation of the "medical side of the operation." Gesse confined his own presentation to the "organizational side"—namely, the availability of blood and, hence, donors.

Gesse stated that the organization of blood transfusions would necessarily differ in war and in peacetime. During wartime, he suggested, Soviet "sanitary organizations" should follow the U.S. example and simply draft "special brigades of donors" into service, the same way soldiers were drafted into the army. During peacetime, however, such conscription, Gesse felt, could not be justified. Gesse surveyed the existing systems of voluntary donorship in Germany, Austria, and Switzerland, as well as his own experience with the organization of volunteers recruited from among the students and personnel of his clinic. He found the system of volunteers inefficient and insufficient.

Gesse thought that the only solution was to organize professional paid donors as in the United States, where "donors even have their own trade union." He suggested that the department establish a "central institution of paid donors," which would serve all hospitals in the city. He outlined a preliminary organizational structure, personnel, and budget for such an institution, including a detailed scheme of payments for donors. He even sketched out certain "legal problems" related to the proposed institution, "to the extent they might be addressed by a physician," focusing in particular on the subject of harm, which might result from a blood transfusion, to both donor and recipient. "Undoubtedly, the interests of the population demand urgent action," Gesse concluded, "and any delay in the organization of paid donorship could deny us the possibility of saving a human life with a simple procedure—the transfusion of blood."

The meeting participants obviously agreed with Gesse's passionate report. They adopted a long resolution that endorsed Gesse's proposals in their entirety.[21] There remained, however, a "small" hurdle. "Since there are no legal guidelines for the question [of paid donorship] and [the meeting] cannot produce any definitive instructions," the resolution declared, "it is necessary to call forth a special meeting of jurists to draft the legal side of the question and to present the draft for approval to [state] agencies in charge."

A few days later, at the Eighteenth Congress of Russian Surgeons held in Moscow from May 27 to 30, Gesse tried to rally support for his idea. On the first day, Gesse's assistant Iosif Maiants delivered to a special section of the congress his report "on indications for and results of blood transfusions" performed at Gesse's clinic. During the discussion that followed Maiants's presentation, Gesse again suggested organizing "professional donorship as in England and America."[22] But this time, his proposal did not meet with universal approval. Some surgeons felt that donorship could not be treated as a profession. The Odessa transfusiologist Barinshtein relayed his own "quite satisfactory" experience with a regular group of twenty volunteer donors he had recruited from among the students at his clinic. The section's chairman, eminent surgeon Genrikh Turner, suggested that donorship could be "viewed as a manifestation of altruism, of familial relations, or even as selling a part of one's body," but not as a profession. As a result, Gesse failed to prod the congress into passing a special resolution on the issue.

But he did not give up. In the July–August 1926 issue of the *Leningrad Medical Journal*, Gesse published the full text of his report to the May meeting in Leningrad, along with the meeting's resolution. Immediately after Gesse's contribution, however, the journal carried an article by Lev Dembo (1926), a well-known jurist who had published extensively on various legal issues in medical practice. Entitled "Judicial Prerequisites to the Question of Paid Donorship," the article addressed numerous, and seemingly insurmountable, legal issues related to blood transfusions.

Soviet newspapers and magazines had regularly described the U.S. system of "paid donors," "donor unions," and "exorbitant prices" for blood as an example of the ultimate exploitation of "poor classes" by the "blood-sucking" (literally) capitalists.[23] So the commercialized selling and buying of human blood—widely accepted in "capitalist" countries—clearly undermined the very foundations of the "socialist" state and its legal system. Dembo tried to circumvent this contradiction by suggesting that the donor could be seen as a worker who sells certain services (and not a part of his body) to the hospital, which under the conditions of NEP could be deemed acceptable. He even deployed an analogy with wet nurses, who, he alleged, also sold their "services."[24] Yet, in the end, he admitted that the "legal problems arising in relation to the institution of donorship are so new and complicated that for their correct resolution, it is necessary to coordinate all the required regulations among the Commissariats of Labor, Justice, and Social Security," or perhaps even to issue an entirely new set of laws.

Alas, such regulations and laws were not forthcoming. Each surgeon, and each hospital practicing blood transfusions, was left to their own devices. Some used volunteers, others paid for blood out of funds they had extracted from Narkomzdrav's central and regional offices, and still others used a combination of paid and unpaid donors. But the country's transfusiologists all felt that the situation was unsatisfactory. Even Bogdanov, who had apparently had no shortage of volunteers from among his students at various "Communist" educational institutions, realized the need for an organized system of donorship after he had moved his operations from Maloletkov's apartment to the new institute and scaled up blood transfusions from "about ten" to "more than two hundred" per year. The institute had to develop a network of donors, both paid and unpaid, which, by the time of Bogdanov's death, had grown to more than four hundred individuals, to serve its ever-expanding clientele suffering from "Soviet exhaustion."[25] But this network could not even begin to serve the needs of numerous Moscow hospitals. For a city of more than two million inhabitants, it was literally a drop (of blood) in a sea (of need).

Russian transfusiologists regularly raised the issue of donorship at various professional meetings. Beginning with Gesse's 1926 presentation, practically every congress of Russian surgeons—held annually during the 1920s—addressed the issue. So did congresses of Ukrainian surgeons. So too did frequent meetings of the Pirogov Society of Russian Surgeons in Leningrad and of the Russian Surgical Society in Moscow.[26] But all their pleas for instituting a system of donorship went unanswered. This certainly prompted Russian transfusiologists to search for alternatives. In 1927, Shamov and his students began experimenting with transfusions of blood taken from cadavers (Kostiukov 1927; Shamov and Kostiukov 1929).[27] Other researchers focused their efforts on developing methods of conserving and storing blood for long periods of time (e.g., Belen'kii 1930; Arutiunian 1932; Braitsev 1932; Balakhovskii et al. 1932).

Perhaps the situation would never have changed, if it were not for a new revolution, the "revolution from above."[28] During the late 1920s, Stalin began to consolidate his personal power over the Bolshevik Party and that of the party apparatus over the nation. The "Great Break," as Stalin named it, marked drastic changes in all facets of life, including the abolition of the NEP, the collectivization of the peasantry, crash industrialization, the launching of the ambitious first Five-Year Plan, and, most important for our story, extensive militarization. It also resulted in the replacement of practically all commissars in all state agencies with trusted Stalinists. In early 1930, Semashko was dismissed from his post, and the entire system of medical education, research, and services was reorganized. Along with medical science and scientists, Soviet physicians were "mobilized to the service of socialist construction."[29]

Although as early as 1923, Kukoverov had urged the Narkomzdrav Main Military-Sanitary Directorate to make blood transfusions a priority, directorate officials, following the verdict of the Medical Scientific Council, were slow in acknowledging the possible military importance of the procedure. Moreover, despite the fact that the country's first group of transfusiologists emerged within the Military-Medical Academy, the connections between their research and the actual medical services in the Red Army remained practically nonexistent.

The situation began to change during the late 1920s as the country was undergoing wide-ranging militarization.[30] In 1929, the official journal of Soviet military medicine, *Military-Sanitary Affairs,* started to carry frequent articles on blood transfusions to popularize the technique among military doctors, and quite a few of those articles were written by Bogdanov's former "comrades-in-blood-exchanges," Maloletkov (1929) and Gudim-Levkovich (1929a, 1929b), as well as Elanskii (1931). The mainstream medical periodicals began publishing articles on the uses of blood transfusions in "future military conflicts" (e.g., Spasokukotskii 1931). Elanskii (1929) wrote one of the first detailed articles on the subject of "blood transfusions under conditions of military combat" for the *New Surgical Archive.* Furthermore, in May 1930, the Medical-Sanitary Administration of the Red Army issued a list of research topics considered of special importance for the army. The list was forwarded to the Narkomzdrav Scientific Medical Council, which in turn distributed it among subordinate research institutions. Number 20 on the list was "blood transfusion."[31]

The recognition of the military importance of blood transfusions provided Soviet physicians with weighty arguments in their dealings with patrons in state and party agencies on the questions of building special institutes for blood research and creating a system of blood services in the country. Once again, Shamov played a pivotal role in this process. A few years earlier, in 1926, he had managed to persuade the Ukrainian Narkomzdrav to support the establishment of the Permanent Commission to Study Blood Groups and its bulletin mentioned above, but he had failed to convince state officials of the need to create a special institute that would provide a solid institutional base for the commission, as well as for Shamov's own research on blood transfusions. Yet he did not give up on the idea. Furthermore, his vision extended beyond founding an institute to pursue his own favorite research subject. As with his commission for the study of blood groups, Shamov envisioned a country-wide system of interconnected institutions for blood transfusions.

In September 1930, the Fourth Congress of Ukrainian Surgeons was to convene in Kharkov. Although in name this was a forum for surgeons working in the Ukrainian Soviet Socialist Republic, in reality practitioners from all over the Soviet Union regularly attended the conference. Shamov used his position as a leading member of the congress's organizing committee to ensure that the subject of blood transfusions would be the congress's first plenary topic. In preparation for the congress, he sent out a detailed questionnaire to canvass the opinions of practically all Soviet transfusiologists. In his keynote address to the congress, Shamov (1931) detailed the "sorry state" of the operation in the country: starting with his own first transfusion in 1919, only 3,995 blood transfusions in toto had been performed during eleven years in the entire country, while the Mayo Clinic alone performed nearly twice as many annually. Shamov found this situation unacceptable. Congress participants agreed. Out of the fifteen reports delivered by leading transfusiologists—including Bagdasarov, Barinshtein, Bogomolets, Elanskii, Kramarenko, and Vlados—that followed Shamov's opening address, only one-third dealt with concrete research and clinical questions. Another third addressed the uses of blood transfusions in the military and the last third the issue of donors. The congress established special working groups to address three main problems facing Soviet transfusiology—organization of donorship, legal issues in blood transfusions, and transfusions under conditions of military combat—and to prepare corresponding resolutions. Approved by the congress, the resolutions were forwarded to the authorities.

The barrage of documents sent by the country's leading surgeons to the highest government offices made the desired impression. In December 1930, the Ukrainian Narkomzdrav established the Ukrainian Institute of Blood Transfusion in Kharkov under Shamov's directorship. A few months later, Elanskii founded a "central station" for blood transfusions in Leningrad, and the next year, he "converted" the station into a full-blown institute of blood transfusion.

In accordance with the general policy of centralized control of everything and anything put in effect during the "revolution from above," the most important developments took place in Moscow at the Bogdanov Institute. In 1930, Bogomolets had been appointed president of the Ukrainian Academy of Sciences, and early the next year, he left Moscow for Kiev. His deputy for "administrative affairs," Bagdasarov, became the institute's director. Almost immediately, things changed. The institute began publishing its own journal, *Modern Issues in Blood Transfusions and Hematology,* and was renamed the *Central* State Institute of Clinical and Experimental Hematology and Blood Transfusions. The name change clearly reflected the institute's new mission: in addition to conducting blood research, the institute was now to spearhead and coordinate the creation of a countrywide system of institutions, personnel, and technologies for introducing blood transfusions into the daily practice of Soviet physicians, above all, in the military services.

In just five years, from 1931 to 1936, under Bagdasarov's capable administration, such a system came into being.[32] Bagdasarov helped organize 49 "branches" of his institute throughout the country, many of which quickly grew into separate institutes for blood transfusions (in Minsk, Kazan, Voronezh, Gorky, Sverdlovsk, Ivanovo, and other cities). The institute's branches in turn fostered the establishment of 514 "basic stations" for blood transfusions in the neighboring regions and rural districts. The budget of these institutions increased tenfold in the same period. The total number of transfusions performed in the country soared from 2,400 in 1932 to 48,790 in 1936.

As in many other countries around the world, this incredible growth reflected the emergence of a system of organized donorship, even though the Soviet system differed in many respects from those created elsewhere.[33] During the 1930s, the Soviet Red Cross Society, which became a de facto arm of the Red Army medical services, waged a huge propaganda and mobilization campaign to recruit volunteers, hailing blood donations as the "honorable duty of every Soviet citizen." The Young Communist League (Komsomol) joined the effort, as did other "voluntary" organizations such as the USSR Union of Societies for the Assistance to Defense, Aviation, and Chemical Industry.[34] As one of the propaganda posters

БЫТЬ ДОНОРОМ—
ВЕЛИКАЯ ЧЕСТЬ ДЛЯ ПАТРИОТА!

6.2 A propaganda poster for blood donations, 1941. The caption
reads, "To be a donor is a great honor for the patriot!"

issued by the Red Cross Society proclaimed, "To be a donor is a great
honor for the patriot!" The campaign culminated in a special decree
issued by SNK in April 1935, "On the Cadres of Donors." The decree affirmed the principle of voluntary donorship: there could be no "selling"
of blood and no "professional" donors in a socialist society. It defined
donorship as "an especially useful social function and a voluntary act,"
which all citizens should perform "without leaving their main job." But
the decree did introduce "monetary compensation" for donors' "extra
nourishment," along with short-term paid leaves from work.[35]

By the mid-1930s, then, blood transfusion had become a standard medical procedure in Soviet Russia. From 1930 to 1936, with the Bogdanov Institute in the lead, the country developed the necessary infrastructure, including research institutes, factories producing equipment, "base stations," and the cadres of donors, to create a statewide system of blood services, along with a system of collecting and storing blood. These developments led to several important innovations, such as the use of cadavers' blood, new technologies of conserving and transporting blood, and the establishment of the world's first "blood banks," which won Soviet transfusiologists the respect of the international community, clearly manifested during the first two international congresses on blood transfusions in Rome (1935) and Paris (1937) (Schmidt 2009).

But all these achievements came about after Bogdanov's death. So, what did Bogdanov actually contribute to the development of Soviet transfusiology, and biomedical sciences more generally? Why after his death were his ideas of blood exchanges so rapidly transformed from "a principally new" and "proven in the clinic" direction of blood research into "medieval mysticism"? And why did his "tectological" theory of senescence fail to gain support and acceptance among Soviet biologists and physicians?

7 Lessons and Legacies

A MARTIAN ON EARTH

Notwithstanding his avowed collectivism, in his life, work, and even death, Alexander Bogdanov was a truly singular phenomenon: a "Martian stranded on Earth," as he disguised himself in a poem that sketched out a third novel of his "Martian" fictions.[1] Judging from the jottings scattered through Bogdanov's notebooks, the novel was intended to be a "synthetic" mirror image of both *Red Star* and *Engineer Menni*: a chronicle of the Russian Revolution and its aftermath seen through the eyes of another "naïve" observer, this time traveling in the opposite direction, from Mars to Earth.[2] It would have been a great source for understanding how Bogdanov himself perceived the earthly realization of what he had envisioned in his previous Martian novels: how did his vision of physiological collectivism, his Marxist analysis of what science ought to be in the society of a victorious socialist revolution, and his projections about what biological and psychological changes a socialist revolution bestowed upon its victors and victims fare in Soviet realities? Alas, Bogdanov did not live to write that novel.

The "trinity" of Bogdanov's persona—a revolutionary dreamer, an SF writer, and a visionary biologist at one and the same time—certainly made him unique in more than one respect. What meaningful comparisons could be made with, and what general conclusions could be drawn from, a singularity, a Martian among the Earthlings? Yet, if we follow Bogdanov's "tectological" thinking, even the life of an alien must conform to, and hence reflect, certain "general principles" operating within the larger system in which that alien exists. Otherwise, the alien cannot possibly survive. Indeed, Bogdanov's death has often been construed in exactly this way. Bogdanov died in April 1928, just as the "revolution from above" began to, once again, reshape dramatically the political, institutional, and cultural terrains of Soviet Russia. In his eulogy at Bogdanov's grave, Nikolai Bukharin told his audience that, together with several other old friends of the deceased, he had come to the gathering directly from a plenary meeting of the Central Committee. What he did not tell his audience, but what they would learn in a few days' time from the newspapers, was that the Central Committee had been discussing the so-called Shakhty affair—a forthcoming show trial of several "bourgeois" engineers accused of wrecking the coal mines entrusted to their management. The Shakhty affair was one of the first signs of Stalin's "Great Break" in the life of the country, which would cost millions of lives and would lead to the establishment of Stalin's dictatorial rule. With foresight of what was to come—purges, Gulag, execution squads—some commentators have assumed that Bogdanov believed that he could not possibly survive in the new, Stalinist Russia, having barely survived his arrest in the previous, Bolshevik Russia. This assumption is the ultimate source of the recurrent "theory" that Bogdanov's death was a suicide, or perhaps even a homicide. By all available evidence, it was neither.

Yet Bogdanov's life and death, "outlandish" and "Martian" as they might seem, do reflect certain "organizational principles" of the Soviet system, and especially the Soviet science system, and thus provide us with some general observations on the interplay among the three revolutions: the Bolshevik Revolution, the "big science" revolution, and the experimental revolution in the life sciences. They offer new insights into interactions between Russian scientists and their Bolshevik patrons, between theoretical visions and scientific practices, and between societal values and moral economies of scientific research. They also illuminate the profound influence that Bogdanov's Marxist ideas of what science ought to be exerted both on his own scientific pursuits and on Soviet science as a whole.

SCIENTISTS AND PATRONS

Party patrons and Russian scientists did not necessarily share the same interests, even if the ultimate result of their interactions was one and the same: the establishment of large research institutions. But as long as each partner *thought* that they got what they wanted from the joint enterprise, the collaboration continued, no matter what those outside of each particular relationship did or said about it.

Contrary to the recurrent myth that Bogdanov initiated, and actively lobbied party and state patrons for, the creation of an institute of blood transfusion (e.g., Belitskaia 1974, 74; Afonin 2002, 45), Bogdanov's appointment was but a historic accident. Just as Stalin, Trotsky, Rykov, and other members of the Politburo were pondering the health of the country's ruling elite, Bogdanov had "cured" Krasin, a member of that elite, with a blood transfusion. Prior to this incident, Bogdanov had worked privately and had shown no intention of spearheading the institutional and practical development of blood research and blood services in the country or even of joining its ever-growing community of transfusiologists. In early 1921, Bogdanov did endeavor to air his ideas of blood exchanges among professionals: a trusted comrade passed a copy of his 1910 program on the "transplantation of blood" to a certain "esteemed colleague" in Petrograd (perhaps to Shamov, who had just published his first article on blood transfusions). Bogdanov was told that the "colleague" had simply dismissed his opus. Not discouraged, Bogdanov sent the "colleague" a long letter, asking for a face-to-face meeting to discuss his current ideas, which, he assured his correspondent, differed substantially from the early "raw" text and now incorporated all the latest advances of modern science.[3] As far as I have been able to ascertain, Bogdanov never received a response, and, apparently offended by such a reaction, he never again tried to contact any of his "brothers-in-blood-transfusions," sharing his thoughts and experiments only with a close group of comrades.

"Curing" Krasin attracted the attention of the Bolshevik Party leadership and propelled Bogdanov onto the professional and public scene in the spring of 1926. The party bosses overruled the opposition of the Narkomzdrav Scientific Medical Council and created an institute of blood transfusion, thus furthering the institutionalization of the discipline of hematology, on the assumption that Bogdanov's work could help "protect" the health of the country's ruling elite, as it helped Krasin (even if not for too long). This was certainly a far from unique happenstance: as I have detailed elsewhere, similar concerns about the health of a top-level Bolshevik (Lenin's wife, Nadezhda Krupskaia, who suffered from hyperthyroidism) secured the support of party patrons at the early stages in the institutionalization of Soviet endocrinology (Krementsov 2008).

Bogdanov's *personal* interest in blood transfusions derived from his vision of "physiological collectivism" that would unite socialist humanity and extend the life span of its members. Like many other intellectuals of the time, he was excited and inspired by the seemingly endless possibilities that concurrent advances in experimental biology and experimental medicine were opening up. The transformation of the "comradely exchange of life" into the "grafting of blood" and then into the "tectological" theory of senescence shows that Bogdanov followed attentively the "visionary biology" of the 1910s and 1920s. He was constantly readjusting his own vision to the newest developments in uncovering the "secrets of life and death." And he was certainly not alone in this respect: Helán Jaworski in France developed similar ideas, though different techniques, during exactly the same period.

Certain elements of Bogdanov's vision reflected his long involvement both in Marxist polemics and in more general Russian debates over the interrelation of the "individual" and the "collective."[4] His "physiological collectivism" perhaps appealed to some of his old comrades in the Bolshevik Party, as it did to some of his young communist students—after all, several of them took part in the experiments with blood exchanges. The numerous editions of *Red Star* issued by various publishers after the revolution probably also helped disseminate Bogdanov's ideas about blood exchanges, particularly among younger readers. But his vision of a new, "physiologically united" humanity, popular as it might have been among some enthusiasts, had very little (if anything) to do with the needs of the party apparatus stricken by an epidemic of "Soviet exhaustion." Nor did it have much relevance to the concerns of Russian physicians who urged the introduction of blood transfusions into clinical practice and the creation of a countrywide system of blood services, but were impeded by the dubious legal position of donorship in the newborn socialist state. Not surprisingly, with the death of its author, Bogdanov's vision "died" too. Indeed, just a few months after his death, the Society of Materialist Physicians held a special session devoted to "The Critique of the Theory of 'Physiological Collectivism'" (Veil' 1929).

Richard Stites (1989), in his thoughtful analysis of utopian visions in 1920s Russia, has argued convincingly that "revolutionary dreams" withered away and perished as a result of the ideological strictures imposed on Soviet cultural and social life by the "revolution from above." The timing of the "death" of Bogdanov's vision seemingly supports this argument. But there was more to the "death" of Bogdanov's ideas of the "struggle for viability" than their "utopianism," or their irrelevance to the actual needs of the party leadership and to the aspirations of the medical community. After all, research on other projects of visionary biology, including various means of "rejuvenation" such as blood "purification," gland transplants, and hormone supplements, continued unabated (and generously supported by the state and party agencies) through the 1930s and beyond.[5]

VISIONS AND REALITIES

So why were Bogdanov's ideas of blood exchanges so quickly rebranded from "a principally new" direction of blood research into "medieval mysticism"? The reviews of Bogdanov's *Struggle for Viability* published in professional periodicals during his lifetime, along with Bogomolets's article published a year after Bogdanov's death, offer an answer: it was not so much Bogdanov's vision as his *science* that did not correspond to the current standards of the scientific/medical community.[6]

A major reason for this "imbalance," using Bogdanov's term, was a gap that separated Bogdanov's ideas about proletarian science—or, more precisely, about what science is and ought to be in a socialist society—which he had consistently tried to implement in his own work, and the realities of concrete scientific practices of the day. In his numerous musings about science, Bogdanov treated his subject in a peculiar and often self-contradictory way. In part, this peculiarity had its origin in the language. In Russian, the word for "science" is *nauka,* and, like its German counterpart *Wissenschaft, nauka* means a systematic pursuit of knowledge in any and every possible area. So Bogdanov's notion of science included—and was based upon—not just natural sciences but also philosophy, social sciences, and the humanities.[7] Furthermore, Bogdanov's concept was deeply rooted in his personal "scientific" experiences. Prior to his first experiments with blood exchanges, he himself had hardly ever actually done any natural science (other than perhaps repeating some simple experiments in physics and chemistry courses he had taken during his student years). As a result, his notion of science was heavily influenced by his own practical experience—namely, reading, thinking, and writing—in economics, psychology, philosophy, sociology, and literature.

Accordingly, his analysis focused almost exclusively on what could be called a *theoretical*—as opposed to investigative and social—practice in systematic, scientific pursuits of knowledge.[8] Despite his repeated statements that science was an "outgrowth of labor," and especially "practical labor activities" (e.g., Bogdanov 1918e, 8; 1918f, 36), Bogdanov treated science primarily as a *system of knowledge* and concentrated on analyzing its theoretical, epistemological side: ways of knowing, types of explanation and proof, principles of reasoning, and forms of presentation. Forgetting his own Marxist analysis, he did not regard science as a "labor process" that involves not simply *reasoning* about, but actually *doing,* certain things. The questions of how actual scientific research is carried out and how the way one "does" science affects its results never attracted his attention.

As Bogdanov himself candidly admitted in his 1921 letter to the "esteemed colleague," he was "primarily a man of general conceptions, who has under his belt far too little of any laboratory, and more generally, technical practice."[9] Even though in *Struggle for Viability* he envisioned several hypothetical "paths of investigating" blood exchanges, Bogdanov knew next to nothing about how exactly to design and conduct laboratory experiments and clinical investigations, how to create and maintain standard "protocols" for his experiments and records of their results, or how to choose and/or construct special instruments and tools suitable for his research. Characteristically, he never even mentioned control experiments. Although he often pontificated on the importance of mathematics as a universal way to "generalize" knowledge and of statistics as a specialized tool of "inductive generalizations," in his concrete research he utilized neither, despite the fact that in 1925 the Communist Academy issued a special volume entitled *Statistical Method in Scientific Investigations,* which included articles on the application of statistics in biological and clinical research (Smit and Timiriazev 1925). Elanskii's review of *Struggle for Viability* strongly implied that not just Bogdanov's theoretical ideas of rejuvenation by blood exchanges, but the very way he had actually conducted his research, belonged to the fifteenth century, not the twentieth. This lack of expertise in (and understanding of) concrete research practices in contemporary experimental medicine and biology was also probably a main reason for the dismissal of Maloletkov (who, as a practicing physician, also had no research experience) and the appointment of Bogomolets—by that time a reputable *researcher,* as well as a theoretician—as Bogdanov's deputy, and a needed leader, for "scientific affairs" at the Institute of Blood Transfusion.

On the other hand, despite his countless declarations that science was a "collective activity," Bogdanov did not examine closely the possible implications of this view. For him, if a "collective" were presented with all of the necessary and sufficient data, there would be no reason for disagreements and debate: every "rational" human being (possessing of course the same class consciousness) will interpret the data the same way and will come to the same conclusion.[10] In his depiction of decision making on socialist Mars, he vividly portrayed exactly this way of resolving contradictions and differences of opinion during the discussion of a possible Martian invasion of Earth and the subsequent extermination of its inhabitants. Even though Bogdanov himself had collaborated with several of his fellow Marxists in writing textbooks on political economy, he had never been exposed to a collective way of doing science in a laboratory or a clinic, which became a characteristic feature of the "big science" emerging during exactly this time. He had no idea about various forms of social dynamics, which are embedded into the collective pursuit of scientific knowledge and which Robert E. Kohler (1994) has fittingly termed the "moral economy of laboratory life."

Bogdanov did maintain a certain moral economy in his institute, but this economy pertained not to its functions as a research establishment as such but to the more general sense of its being a "collective." For Bogdanov, a collective was a voluntary association of comrades—like-minded, equal in every way, sharing the same goals, ideals, labors, and responsibilities. This is what his "organization of physiological collectivism" actually was, and this is how it indeed functioned. The moment this "band of comrades" expanded into a much bigger collective with the establishment of the Institute of Blood Transfusion, the troubles began. The previous diffuse ("structurally disorganized," to use Bogdanov's own term) circle of comrades was transformed into a formal, structured, and regulated entity, with a particular internal distribution of powers, labors, and responsibilities, which, as in any other institution, had to be observed and maintained. Despite its very privileged position within Narkomzdrav, Bogdanov's institute also became a part of this larger structure and was subordinate to its Administrative-Financial Directorate, its party cell, and the commissar personally. Yet Bogdanov tried to run his institute as if it were still a circle of comrades.

It was Bogdanov's "comradely" managerial style as the director that apparently encouraged his deputies for "administrative affairs"—first Zeilidzon and then Ramonov—to try to seize his directorial powers (such as, in the case of Zeilidzon, the hiring of personnel, and in the case of Ramonov, the disbursement of funds) for themselves. Both Zeilidzon and Ramonov were old party hands, who, unlike Bogdanov, had stayed with the Bolsheviks through the revolution, the civil war, and the perpetual struggles with various "oppositions." This experience, as Bogdanov astutely observed more than once, instilled authoritarianism into the very soul of Bolshevism and killed off the sense of comradeship that had imbued the early workers' circles from which the Russian Social-Democratic Labor Party itself had sprung. As he intimated in a letter to his old comrade Lunacharskii, this authoritarianism and the substitution of "drunken camaraderie" for true comradeship were the major reasons Bogdanov had left the Bolshevik Party in the first place and refused to rejoin it after 1917.[11] But for both of his deputies, this was obviously the normal state of affairs. In both cases, Bogdanov failed to "convert" his deputies to his "collectivist," "comradely" way of running the institute; they both probably perceived it simply as a weakness on the director's part. In both cases, constricted by the formal structures and regulations, Bogdanov could not simply call up a general meeting of his comrades and expel the "unsuitable" members who had refused to accept the moral economy of comradeship he tried to uphold in his institute; in each case, he had to appeal to his patrons (Semashko and Stalin) to get rid of them.

One could argue that Bogdanov's conflicts with his deputies reflected more than a clash of personalities and of two different moral stances. In his penetrating analysis of the moral economy of the most famous research laboratory in Russia, Daniel P. Todes (2002) has demonstrated convincingly that Ivan Pavlov's authoritarian style affected profoundly the production of knowledge and the allocation of credit in his laboratory. Todes has also observed that Pavlov's personal style appeared preadapted to and highly effective under the conditions of a large, ever-growing research enterprise that characterized the transition from the "little science" of the end of the nineteenth century to the "big science" of the twentieth century, and particularly to the Soviet style of "big science." Continuing along this line of reasoning, one can suggest that Bogdanov's personal style actually hindered the development of his institute into a true national center of research on blood transfusions. Bogdanov's managerial style was likely responsible for the institute's failure to carry out its assigned tasks, including the organization of courses for practicing doctors and the development of apparatus and manuals for blood transfusions, despite all the resources its patrons lavished upon the institute. This is why, even though the Narkomzdrav officials approved Bogdanov's report in March 1928, they insisted that he had to "integrate the work of the institute's experimental and clinical sections" and to "expand the institute's contacts with other medical institutions."[12] One could speculate that if Bogdanov had not died in 1928, he probably would not have lasted long as the director, certainly not during the institute's tremendous expansion in the 1930s. This required a man of Bagdasarov's managerial talents.

CONVENTIONS AND CONTENTIONS

The collective character of scientific activities, which Bogdanov so frequently praised, had certain implications that transcended the walls of his institute, but which Bogdanov also failed to recognize. Particular norms, conventions, rules, customs, and mores govern the interactions not only within the smaller collectives of particular laboratories, institutes, and clinics but also within larger collectives as well—from a specific disciplinary community to a national community to the world scientific community as a whole. These social dynamics regulate a variety of "labor activities" performed by members of a particular community. One such activity, in which Bogdanov actively partook, is publishing the results of one's scientific pursuits.

Bogdanov certainly hoped that his readers would perceive *Struggle for Viability* as a scientific treatise, even if written with a general audience in mind. But in many ways his book did not conform to the established norms of scientific publication either in experimental biology or in medicine. Contrary to the repeated emphasis on the profound difference between "faith and science" in his writings on proletarian science (e.g., Bogdanov 1910, 1913b, 1918f), Bogdanov obviously expected his readers to take much in his latest treatise on faith. Although he claimed that his "theoretical ruminations" were "based on a considerable mass of experience accumulated by various sciences; [and on] voluminous and significant data of scientific-practical experiences in Europe and America supportive of our conclusions"(Bogdanov 1927a, 152), he was very "economical" in substantiating this claim with references to actual publications, either on blood transfusions or on senescence. By the start of the twentieth century, exact referencing of works published by other researchers on the subject under investigation had become a standard in scientific publications, an indispensable part of the disciplinary consensus and conventions regarding methods, subjects, and objectives of studies. To give but one example, both Bruskin's and Elanskii's monographs reviewed by Bogdanov contained extensive bibliographies of sources they had used in their own works. But Bogdanov apparently regarded such conventions as unnecessary trappings of "bourgeois" science invented to maintain the exclusivity of the scientific profession and to keep the proletariat in the dark. Furthermore, his own "scientific" experience could provide him with no guidance in this respect, since conventions of referencing in philosophy, economics, and psychology of the time differed substantially from those in experimental biology and medicine. In terms of references, Bogdanov wrote *Struggle for Viability* in exactly the same manner he had written his *Empiriomonism* or *Tectology*, providing only an occasional footnote to a monograph he considered important.

Not only did his book include no bibliography, it contained not a single reference to numerous works on both senescence and blood transfusions that had appeared in Russian professional journals or proceedings of professional conferences. Indeed, aside from Bruskin and Elanskii, whose works he reviewed in the appendix, he referred by name to only five other Russian scientists: Kliment Timiriazev, Nikolai Kol'tsov, Boris Zavadovskii, Konstantin Gess-de-Kal've, and the biophysicist Petr Lazarev; and only the first three were mentioned more than once. For Bogdanov, Timiriazev was certainly a major authority on general biological knowledge, particularly on questions of heredity and evolution. He referred regularly to "Historical Method in Biology," which Timiriazev had revised and updated with some "Marxist" formulations just before his death and which had appeared in book format in 1922. On the other hand, although Bogdanov did use Kol'tsov's and Zavadovskii's literature surveys as a source of information on Steinach and Voronoff's operations, aside from Zavadovskii's critique of his own ideas, he either ignored or dismissed outright their evaluations of various concepts of senescence and techniques of rejuvenation.

Even though Bogdanov was much more generous in his references to Western literature, Keynes's 1922 book remained his major source on all questions related to blood transfusions. He supplemented it with two somewhat newer books on the subject published in 1924 and 1925 in France.[13] He also used available Russian translations of several Western publications, including Dupuy de Frenelle's booklet on blood transfusions and Kammerer's textbook on general biology. But it seems that Bogdanov's main source of information on new advancements in experimental biology and medicine were popular-science magazines, not professional periodicals or monographs. The notebooks of 1913–25, preserved among Bogdanov's personal papers, contain copious notes from various articles published in *Priroda* (Nature), the major popular-science journal in Russia, but nothing from professional medical or biological journals.[14]

In his writings on blood transfusions, Bogdanov always emphasized that prior to the establishment of his institute, the procedure had been virtually unknown in Russia. Yet, in five years, from 1921 to 1926, the country's major medical journals carried no fewer than forty articles on the subject of blood groups alone (Rafal'kes 1926). Judging from the bibliographies in Bruskin's, Elanskii's, and Barinshtein's monographs, they published at least twice that number of articles on blood transfusions per se.[15] None of these articles was even mentioned in Bogdanov's book, to say nothing of the extensive Russian literature on senescence and rejuvenation. He did not even refer to a survey of literature on transfusions with citrated blood published by the member of his own circle, Gudim-Levkovich (1926). A book that included no references to and no discussion of publications by other scientists, particularly compatriots, working on the same subject could hardly be accepted as a scientific monograph. Indeed, exactly at the time of Bogdanov's work on *Struggle for Viability,* Russian physicians were calling for strict enforcement of this particular convention. As one of them put it, "Every author, writing in Russian for a Russian publication, *must* include into his bibliography at least the newest, covering the last four to five years, Russian literature on the subject under investigation" (Rafal'kes 1927,361–63; italics in the original). He even proposed that journals reject all articles submitted for publication without an "obligatory Russian bibliography," and some editors endorsed his proposal. Bogdanov's book clearly failed to conform to this convention.

The "generalized and simplified" language Bogdanov had devised for his tectological analyses and widely deployed in his book also likely contributed to the failure of Bogdanov's concept to win the attention of its readers among professionals. By refusing to employ the accepted scientific terminology of the time, Bogdanov indicated that his work was outside the purview of professionals. This probably played a role in the total silence that greeted *Struggle for Viability* in Russian *biological* journals, even though it had been written from "a general biological viewpoint" and its main subject—a theory of senescence—commanded close attention of the Russian biological community at the time.

Furthermore, as reviewers of his manifesto pointed out, Bogdanov's style of writing and particularly his tone of polemics made his book look like "a feuilleton," not a scientific treatise (e.g., Breitman 1927, 1142). Admittedly, in just a few years, during and especially after the "revolution from above," this style of polemics—emulating the barbed style of vicious intraparty attacks on various oppositions—would thoroughly penetrate Soviet scientific discussions and publications (Krementsov 1997, 45–53). But in 1927, it was still something quite new, and the majority of Russian scientists and physicians clearly considered it inappropriate for a respectable scientific publication.

SCIENCE AND MARXISM

Bogdanov was one of the first to attempt a Marxist analysis of science as a specific form of human activity. Yet in his own scientific practice, he did not follow the obvious precepts that his analysis prescribed. One particular element of Bogdanov's Marxist concept of science clearly clashed with the reality of his actual scientific work and played a major role in the "death" of his research program on blood exchanges following his own death. Surprising as it may seem, that element was the "practicality" of his research. According to Bogdanov's Marxist postulates, "science is the organized experience of a human society," or more precisely, "organized societal-labor experience," and its main function is "to organize labor" (e.g., Bogdanov 1918f, 6, 36). He never tired of repeating that science serves as "an instrument of the organization of social labor," "a means of the organization of production." This "instrumentalist" view often led Bogdanov to equate science with technology, engineering, and even management, as he did in *Engineer Menni,* characterizing its two main protagonists interchangeably as scientists, engineers, and administrators. This instrumentalist view also led Bogdanov to the conclusion that science did in the past—and ought to in the future—work on the tasks "put forward by labor and production," and, more generally, by the needs of society. In Bogdanov's opinion, in a capitalist class society, science serves only the needs of the oppressor and itself becomes an instrument of oppression. Even though in its insatiable pursuit of profits, capitalism did increasingly employ science for the organization of production, it separated science from labor, allowing "bourgeois" scientists to advocate the principles of "pure science" and of "science for the sake of science." In a socialist classless society, with its universality of labor and with the interests of labor and production coinciding, science must and will serve the needs of society as a whole, as it does in *Red Star*. For Bogdanov, there could be no pure science, nor science for the sake of science, under socialism.

Viewed from this instrumentalist perspective, Bogdanov's science of blood exchanges clearly failed to qualify as an exemplar of proletarian science. It did not address the problems of "labor practice" or the "needs of society" at all. Admittedly, it could be adapted to the need of the party elite for a quick remedy against "Soviet exhaustion and attrition," and that was what got Bogdanov into professional science in the first place and what sustained his unassailable institutional position, despite all the critique his work had elicited from the scientific/medical community. But Bogdanov did not address the burning issues of blood transfusion as a practical clinical procedure that could save countless lives, such as the availability of donors, conservation of blood, and development of blood substitutes, which occupied other transfusiologists and which became the major focus of the Bogdanov Institute after its founder's death. As all the reviewers of Bogdanov's manifesto pointed out, he "had insufficiently elaborated the practical-medical side" of blood transfusions. Ironically, if only he had applied his Marxist analysis of science to his own work, Bogdanov might well have himself predicted the ultimate failure of his research program: after all, blood exchanges had certainly much less immediate practical value than "ordinary" blood transfusions.

What remained of Bogdanov's legacy in Soviet "blood science" was his institute. Under the new conditions created by the "revolution from above" and in Bagdasarov's capable hands, the Bogdanov Institute indeed turned into the fulcrum of the emerging system of blood services. But it was not Bogdanov's vision or his research program that made it so. As in many other countries, it was blood transfusions' perceived military importance that ushered in a new era of rapid growth and advancement in the techniques, instruments, infrastructure, legal and financial support, personnel, and research related to the procedure in Soviet Russia.

If on the Soviet transfusion medicine of the 1930s Bogdanov left primarily an *institutional* footprint, to Soviet science as whole, and particularly the life sciences, he bequeathed a lasting legacy of his Marxist concept of proletarian science. Although, as a result of the ideological orthodoxy enforced during the "revolution from above," in the official discourse of the time Bogdanov's philosophical views were reduced to a pejorative, "Bogdanovism," [16] certain basic elements of his concept of proletarian science were actually incorporated into "Marxism-Leninism-Stalinism." Indeed, Bogdanov's ideas—albeit, of course, without ever mentioning their author—laid a foundation for much of Soviet official science policy in the 1930s and beyond.

This was particularly true of his instrumentalist view of science. The very notion of "mobilizing science to the needs of socialist construction," which became a major slogan of the "revolution from above" and a guiding principle of Soviet science policy for years to come, was but an extension and direct application of Bogdanov's ideas regarding what science is supposed to be in a socialist society. Ironically, even Bogdanov's harshest critics of the 1920s adopted and propagated his ideas a decade later. Nikolai Bukharin's report on "Darwinism and Marxism," delivered to a joint meeting of the USSR Academy of Sciences and the Communist Academy in April 1932, illustrates perfectly this paradoxical situation. According to Bukharin (1932, 30), Darwinism was not merely a "theoretical" concept, explaining biological evolution in materialist terms. It was "a scientific force of production," "zoo- and phyto-engineering," which "is directly connected to the process of material production" and which "transforms agriculture into a science-based branch of socialist industry." Not only the thrust but even the language of this statement clearly echoed Bogdanov's numerous pronouncements on proletarian science, although Bukharin never mentioned his name in the speech. [17]

Bukharin's report enunciated the official point of view and was disseminated through all possible channels. Newspapers published an abridged version. *Socialist Reconstruction and Science,* a new journal established as the oracle of the entire science mobilization campaign, published Bukharin's text in its entirety. The report also appeared as a booklet with the first print run of thirty thousand copies (!), supplemented within a few months by a second, equally large printing. It also served as an introduction to a new 1935 edition of Darwin's *Origin of Species* issued in thirty-five thousand copies!

Indeed, during the 1930s, "practicality" became a major requirement for and a major criterion of evaluation (the main "selection principle," in

Bogdanov's terms) of any and all scientific activities in Soviet Russia. It became a key element of the official "Marxist" view of science and a key component of the "Marxist" rhetoric Soviet scientists employed in their dealings with patrons in the party and state apparatus (Krementsov 1997). In this sense, Marxism—in Bogdanov's particular version—did penetrate the very fabric of Soviet science and exerted a profound influence on its development. To give but one notorious example, "practicality" paved the way for the success of Trofim Lysenko's—utterly practical as its very name made plain—"agrobiology" in the Soviet corridors of power. It secured Lysenko's meteoric career within the Soviet agricultural and biological research system, despite the ever-growing criticism of his "science" by numerous peers. It definitely contributed to the wide following his "agrobiology" generated among the new cohort of researchers entering Soviet life sciences in the 1930s. Moreover, Lysenko's favorite research institution—the so-called shack laboratories, which had been created all over the country and which he tirelessly lauded as the foundation of all his science—was an embodiment of Bogdanov's notion of "simplified and generalized" proletarian science, easily accessible to anyone with (or even without) some "basic" education. I leave it to future historians to discover exactly why and how Soviet medicine (and specifically transfusiology) was able to "outgrow" Bogdanov's own proletarian science of blood exchanges while Soviet biology (and specifically genetics) succumbed to Lysenko's proletarian agrobiology.

Bogdanov's life story puts into sharp relief the particularities of the interplay among the three revolutions that shaped the development of biomedical sciences in interwar Russia. The Bolshevik Revolution paved the way to the "little" revolution, creating very favorable conditions for a rapid transition from "little" to "big" science, with a corresponding proliferation of large, specialized institutions in every field of "visionary biology." Bogdanov's long association with the Bolshevik Party provided him a ready access to state and party patrons who fostered the "little" revolution and secured his appointment as director of a large research establishment. But, placed at the head of the Institute of Blood Transfusion, Bogdanov was caught unawares: his managerial style and his actual scientific practices ill fitted the demands of "big" science. Although his research program on blood exchanges was inspired by and grew naturally out of the "mini" experimental revolution in the life sciences, its justification and particularly its execution were much more suited to the "little" science of the previous century than to the highly differentiated, strictly regulated, and thoroughly conventionalized activities of the "big" Soviet science. In the end, it was this misfit that led to the "death" of Bogdanov's research program in the 1930s. Yet, despite the rancorous condemnation of his philosophical ideas and the complete oblivion of his vision of "comradely exchange of life" in Stalin's Russia, the basic elements of Bogdanov's concept of proletarian science—first enunciated in his SF novels—laid a foundation for much of Soviet science policy for years to come, thus making at least some of his Martian "fairy tales" into Soviet reality.

Abbreviations

ARAN   Arkhiv Rossiiskoi Akademii Nauk (Archive of the Russian Academy of Sciences)

ChiP   *Chelovek i priroda* (Man and Nature), a journal

GARF   Gosudarstvennyi arkhiv Rossiiskoi Federatsii (State Archive of the Russian Federation)

GINZ   Gosudarstvennyi institut narodnogo zdravookhraneniia (State Institute of People's Health Protection)

JAMA   *Journal of the American Medical Association*

JHB   *Journal of the History of Biology*

MMZh   *Moskovskii meditsinskii zhurnal* (Moscow Medical Journal)

Narkomfin   Narodnyi Komissariat Finansov (People's Commissariat of Finance)

Narkompros   Narodnyi Komissariat Prosveshcheniia (People's Commissariat of Enlightenment)

Narkomzdrav   Narodnyi Komissariat Zdravookhraneniia (People's Commissariat of Health Protection)

NEP   novaia ekonomicheskaia politika (new economic policy)

NiT   *Nauka i tekhnika* (Science and Technology), a journal

NKh   *Novaia khirurgiia* (New Surgery), a journal

NKhA   *Novyi khirurgicheskii arkhiv* (New Surgical Archive), a journal

Politburo   politicheskoe biuro (Political Bureau of the Central Committee of the Bolshevik Party)

Proletkult   proletarskaia kul'tura (proletarian culture)

RGASPI   Rossiiskii gosudarstvennyi arkhiv sotsial'no-politicheskoi istorii (Russian State Archive of Social and Political History)

RSFSR   Rossiiskaia Sovetskaia Federativnaia Sotsialisticheskaia Respublika (Russian Soviet Federated Socialist Republic)

SF   science fiction

SNK	Sovet Narodnykh Komissarov (Council of People's Commissars)
STO	Sovet Truda i Oborony (Council of Labor and Defense)
TsKK	Tsentral'naia Kontrol'naia Komissiia (Central Control Commission of the Central Committee of the Bolshevik Party)
VD	*Vrachebnoe delo* (Physicians' Affair), a journal
VF	*Voprosy filosofii* (Issues in Philosophy), a journal
VM	*Vecherniaia Moskva* (Evening Moscow), a newspaper
VMA	Voenno-Meditsinskaia Akademiia (Military Medical Academy)
VMIAB	*Vestnik Mezhdunarodnogo Instituta A. Bogdanova* (Herald of the International A. Bogdanov Institute), a journal
VSM	*Vestnik sovremennoi meditsiny* (Herald of Modern Medicine), a journal
VZ	*Vestnik znaniia* (Herald of Knowledge), a journal
ZhDUV	*Zhurnal dlia usovershenstvovaniia vrachei* (Journal for Doctors' Continuing Education)

Notes

CHAPTER 1

1. Since Mikhail Gorbachev's perestroika, the study of Bogdanov has turned into a veritable "industry." In 1989, a new two-volume edition of Bogdanov's major work, *Tectology,* was reissued in Russia after nearly sixty years of complete oblivion (Bogdanov 1989). The next year, a large collection of Bogdanov's works, both fictional and nonfictional, appeared (Bogdanov 1990). In 1995, Moscow historians issued a three-volume set of his previously unpublished materials from the Communist Party archives (Bordiugov 1995). Three years later, John Biggart, Georgii Gloveli, and Avraham Yassour (1998) compiled a monumental guide to Bogdanov's published and unpublished works, which included a nearly complete bibliography of Bogdanov's writings, a very detailed chronology of Bogdanov's life, and a large bibliography of works about Bogdanov. Needless to say, since 1998, the bibliography of various publications on Bogdanov grew considerably (see, for instance, Biggart, Dudley, and King 1998; Gare 2000; Liubutin 2003; Sadovskii 2003a, 2003b; Gloveli 2003; Greenfield 2006; Prot'ko and Gratsianov 2010; and many others). In 1999, a special "International A. Bogdanov Institute" was established in Ekaterinburg, "as the organizational-scientific center for the unification, coordination, and development of fundamental and applied investigations by the Russian and foreign scientists who use creatively ideas from the scientific legacy of the great Russian thinker Alexander Alexandrovich Bogdanov" ("Otkrytie Mezhdunarodnogo Instituta A. Bogdanova" 2000). The institute almost immediately began publishing its own journal, which carries numerous articles on Bogdanov and his works (see http://www.bogdinst.ru/vestnik).
2. The only attempt I am aware of to find a unity in Bogdanov's diverse activities is Adams 1989.
3. See, for instance, an article entitled "Bogdanov as Scientist and Utopian" (Gloveli 1998), which, despite its title, does not mention Bogdanov's blood research at all.
4. See, for instance, Chistova 1967; Belitskaia 1974; Belova 1974; Huestis 1996, 2002, 2007; Mikhel' 2006; Iagodinskii 2006a; and Donskov and Iagodinskii 2008.
5. See Bogdanov 1984; and Simonov and Figurovskaia 1988.
6. See, for example, Suvin 1971; Jensen 1982; Stites 1984; Graham 1984a, 1984b; Basile 1984; and Greenfield 2006. For a suggestive analysis of Bogdanov's novel in the context of Western science and science fiction, see Adams 1989.

7. In a recent dissertation on the development of SF in fin de siècle Russia, Anindita Banerjee attempted to link Bogdanov's fiction with his science by appealing to "mystical tradition [that] provided a crucial bridge between his utopian vision with [*sic*] actual scientific practice" (Banerjee 2000, 268), which, in my opinion, misses the point completely.

8. See, for example, Diamond 1980; Gunson and Dodsworth 1996; Starr 1998; Giangrande 2000; Moore 2003; Lederer 2008; and many others.

9. See, for example, Farr 1979; Schneider 1983, 1995, 2003; and many others.

10. See, for example, Schlich 1996; Schneider 1997; Pelis 2001b, 2002; and Bauduin 2007.

11. See, for example, Huestis 1996, 2002.

12. See, for instance, Oleinik 1955; Gavrilov 1968; Belova 1983; Afonin 2002; Pertsev 2003; and Mikhel' 2006.

13. Compare, for instance, Tartarin 1994 and Mikhel' 2006.

14. See, for example, Graham 1967, 1990; Esakov 1971; Bailes 1978; Krementsov 1997, 2006; and Tolz 1997.

15. See, for example, Bastrakova 1973; Graham 1975, 1993, 1999; Lubrano and Solomon 1980; Vucinich 1984; Solomon and Hutchinson 1990; Adams 1990; and many others.

16. This "experimental revolution" was clearly manifested in the establishment in the United States of the Society of Experimental Biology and Medicine in 1903 and of the Federation of American Societies for Experimental Biology in 1912. On the early history of the new field, see Gaskins 1970; and Allen 1975.

17. Arguably, it reached its second climax in the 1950s and 1960s, when, following the deciphering of the genetic code, both experimental medicine and experimental biology were transformed into "molecular" medicine and biology.

18. This is readily evident in the title of a famous work by one of the leading proponents of the new field in the United States, Raymond Pearl (1922), *The Biology of Death*. In his excellent monograph, Philip J. Pauly (1987) has explored the general development of these ideas in U.S. contexts, while Jon Turney (1995) has analyzed public responses to these ideas in U.S. and British contexts.

19. See, for instance, Adams 2004.

20. See, for example, the report on the 1921 annual meeting of the Federation of American Societies for Experimental Biology by the federation's executive secretary, Chas. W. Greene (1922) in *Science*.

21. On the role of private foundations in the development of experimental biology and medicine in the United States, see Kohler 1991, 2003.

22. On the Bolsheviks' technocratic and pragmatic attitudes toward science, see Bailes 1978; Krementsov 1997.

23. Stites has thoroughly explored the role of the Bolsheviks' "revolutionary dreams" of a "new society" in various aspects of their social policies, but he has not examined the involvement of Russian science and scientists in both the articulation and the realization of these dreams.

24. For a fascinating examination of these visions through the medium of Soviet cinema, see Widdis 2003.

25. On the history of early Soviet science fiction, see Nudel'man 1966; Britikov 1970; McGuire 1985; Revich 1985; and Banerjee 2000.

26. For an analysis of one such work, Aleksandr Beliaev's *The Head of Professor Dowell*, in its relation to contemporary science, see Krementsov 2009.
27. On the institutional effects, see Esakov 1971; on intellectual influences, see, for instance, Graham 1987; Allen 1991; and Todes 2007.
28. See, for instance, Joravsky 1961; and Krementsov 1997.
29. See, for instance, Lecourt 1977; the French original had been published in 1976 under the title *Lyssenko: Histoire réelle d une science prolétarienne* (Lecourt 1976), with an introduction by France's leading Marxist, Louis Althusser.
30. See, for instance, Mänicke-Gyöngyösi 1982.
31. For a synthesis of the first two sets of arguments, see Todes and Krementsov 2010.
32. There are several other Bolsheviks—for example, the mineralogist Nikolai Fedorovskii and the social hygienist Nikolai Semashko—who, after the revolution, occupied important positions in the Bolshevik agencies in charge of science and at the same time were "practicing" scientists. Unfortunately, there are no detailed studies of either Fedorovskii's or Semashko's science.

CHAPTER 2

1. For analyses of historical developments in blood transfusions in the West, see Diamond 1980; Gunson and Dodsworth 1996; Starr 1998; Giangrande 2000; Lederer 2008; and many others.
2. On Blundell and his work, see Young 1964; and Pelis 1999.
3. For historical analyses of the development of blood transfusions and blood services in Russia, see Oleinik 1955; Gavrilov 1968; Afonin 2002; Pertsev 2003; and Mikhel' 2006.
4. For a detailed account of the early history of blood transfusions in nineteenth-century Russia, see Oleinik 1955, 13–114. Although clearly affected by the ideological and political biases of its time, Oleinik's four-hundred-page monograph contains a wealth of information on the development of blood transfusions in Russia up to the time of its publication.
5. On the use of saline solutions and the disillusionment with blood transfusions in Britain, see Pelis 1997, 2001a.
6. Cited in Oleinik 1955, 104. Borngaupt played on the double meaning of the Russian word *baran*, which means simultaneously "a male lamb" (a preferred source of blood in the early history of blood transfusions) and "a stubborn, narrow-minded person" or "a moron."
7. On blood groups, see Farr 1979; Schneider 1983, 1995; Schlich 1996; Schneider 2003; Bauduin 2007; and many others.
8. On the transfer of the North American expertise in blood transfusions during the Great War, see Schneider 1997; and Pelis 2001b.
9. At the beginning of World War I, in 1914, St. Petersburg was renamed Petrograd; in 1924, it was renamed Leningrad and regained its original name—St. Petersburg—in 1991.
10. For a general overview of the political, economic, and social consequences of the Russian civil war, see Koeniker, Rosenberg, and Suny 1989; for a highly readable account of the war, see Lincoln 1989.

11. Each republic of the Soviet Union established a similar system of governmental agencies in charge of health and medicine, and each republic had its own Narkomzdrav. Until 1936, when the USSR Narkomzdrav was created, the RSFSR Narkomzdrav had played the role of the coordinating center for the Narkomzdravs of all other republics.

12. For a general overview of Narkomzdrav and its activities in the first decade of the Bolshevik rule, see Solomon and Hutchinson 1990.

13. For the early history of this institution, see a special volume issued for its fifth anniversary (Tarasevich and Liubarskii 1924).

14. For a contemporary account of the growth of research institutions under Narkomzdrav in the early 1920s, see Tarasevich and Liubarskii 1924.

15. For a general overview of the NEP, see Fitzpatrick, Rabinowitch, and Stites 1991.

16. See, for instance, an analysis of the disciplinary development in Soviet endocrinology in Krementsov 2008.

17. See the State Archive of the Russian Federation (Gosudarstvennyi arkhiv Rossiiskoi Federatsii; hereafter, GARF), *fond* (collection) A-482, *opis'* (inventory) 25, *delo* (file) 104, *list* (page) 33; hereafter, such references will be given as GARF, f. A-482, op. 25, d. 104, l. 33; see also GARF, f. A-482, op. 25, d. 104, l. 44 rev.; d. 121, ll. 1–3.

18. For the complete text of Tsypkin's proposal, see GARF, f. A-482, op. 25, d. 121, ll. 1–3.

19. GARF, f. A-482, op. 25, d. 104, l. 33.

20. GARF, f. A-482, op. 25, d. 104, l. 44 rev.

21. For the complete text of the memorandum, see GARF, f. A-482, op. 25, d. 306, ll. 1–7. All the following quotations are from this source.

22. For brief biographies of Shamov, see Novachenko 1968; and Nechai 1977. I base my account of Shamov's career on the collection of his personal papers preserved in the Military Medicine Museum of the Russian Federation Ministry of Defense in St. Petersburg. References to this collection will hereafter be given as "Shamov's papers." I am grateful to the museum's staff, particularly the curator of archival collections Igor' Kozyrin, for help in locating the documents and photographs related to the story and for permission to use them in this work.

23. For a biography of Fedorov, see Ivanova 1972.

24. On Crile's work with blood transfusions, see his memoirs, Crile 1947; also English 1980; Hermann 1994; and Pelis 2001b.

25. For historical accounts of Crile's work on blood transfusions see Lederer 2008; and particularly, Nathoo, Lautzenheiser, and Barnett 2009.

26. For Carrel's own description of his technique, see Carrel 1907.

27. Landsteiner 1901; and Moss 1910.

28. Shamov to Cushing, March 28, 1915, Shamov's papers.

29. See *Moskovskii meditsinskii zhurnal* (hereafter, MMZh) 2 (1926): 129.

30. See, for instance, one of the first literature surveys published on the subject in early 1922 in *Gynecology and Obstetrics* (Tsetlin 1922).

31. See, for instance, Shilovtsev 1923; Neiman 1925; and Ushakov 1927.

32. See the meeting's protocol in *Turkestanskii meditsinskii zhurnal* 3, nos. 1–8 (1924): 141–43.

33. For a brief biography of Gesse, see Filatov 1973.
34. For a brief biography of Bruskin, see "Iakov Moiseevich Bruskin" 1969.
35. See reports on this meeting in *Vestnik sovremennoi meditsiny* (hereafter, *VSM*) 5 (1925): 46 and in *MMZh* 5 (1925): 129.
36. See *MMZh* 5 (1925): 128.
37. See the meeting protocol in *Novaia khirurgiia* (hereafter, *NKh*) 4, no. 1 (1926): 45.
38. See the meeting protocol in *NKh* 4, no. 2 (1926): 58–59.
39. See report on the conference in *Novyi khirurgicheskii arkhiv* (hereafter, *NKhA*) 6, nos. 21–22 (1925): 148–50.
40. See announcement for the course in *NKhA* 12, no. 2 (1927): 326.
41. A year later, in 1926, researchers at the Ukrainian Institute of Workers' Medicine published results of their experimental studies on various saline solutions used for transfusion in patients who suffered extensive blood loss (Brikner, Suponitskaia, and Charnyi 1926).
42. See a contemporary survey article on the group's work, Zhitnikov 1927; and Rubashkin 1929.
43. See *Biuleteni postiinoi komisii vivchania krov'ianikh ugrupovani*, or *Verhandlungen der stadigen commission fur blutgruppenforschung* (*Ukrainisches Zentralblatt fur Blutgruppenforschung*), 1927–29.
44. See, for instance, "Perelivanie krovi" 1923; Blinkov 1925; Bruskin 1926; Bakunin 1927; and Vasilevskii 1928b.

CHAPTER 3

1. For the most recent "scientific biography" of Bogdanov, see Iagodinskii 2006b. John Biggart, Georgii Gloveli, and Avraham Yassour (1998, 459–80), in their monumental guide to Bogdanov's published and unpublished works, included a very useful "biographical chronicle" of Bogdanov's life. Much biographical information is contained in Bogdanov's "personnel files" in Narkomzdrav (GARF, f. A-482, op. 42, d. 590) and the Communist Academy (the Archive of the Russian Academy of Sciences [Arkhiv Rossiiskoi Akademii Nauk; hereafter, ARAN], f. 350, op. 3, d. 190, ll. 1–4; and d. 2, ll. 104–8, 110–12).
2. The best available accounts of the Russian populist movement remain Billington 1958; and Wortman 1967. The latter has recently (in 2008) been reprinted in a paperback edition.
3. See, for instance, Ballestrem 1969; and Jensen 1978, 8.
4. Some scholars have argued that Bogdanov was also deeply influenced by Arthur Schopenhauer (Dobronravov 2001). One can also add Wilhelm Wundt and Hermann Helmholtz to the list of scientists/philosophers whose influences could be discerned in Bogdanov's treatise.
5. Although there are quite a few studies of the influence of positivism on Russian philosophy and social sciences, there is not a single book-length examination of Spencer's influence on Russian natural sciences (see Vucinich 1976).
6. On Bogdanov's positivism see Gusev 1995; and Soboleva 2007.
7. Unfortunately, there is not a single book-length examination of the impact of Haeckel and his ideas in Russia. For some suggestive observations, see Vucinich 1988,151–95; Tsalolikhin 1993; and Banerjee 2000,224–26.

135

8. On Haeckel's monism, see Holt 1971; and especially Richards 2008.

9. At that time in Russia, there were two degrees in philosophy, magister and doctor, and it seems likely that Bogdanov was preparing a magister dissertation.

10. For a brief biography of Anfimov, see Kavtaradze 1960. Unfortunately, this short booklet focuses predominantly on the Soviet period of Anfimov's work.

11. See a program of studies written for the students of his department, Anfimov 1915.

12. For Zelenogorskii's life story, see Zelenogorskii 1998, 11–16.

13. See Zelenogorskii 1876. This work, along with several others, has recently been reprinted in Russia; see Zelenogorskii 1998.

14. For a biography of Timiriazev in English, see Platonov 2001.

15. See also Bogdanov's notes for his speech at a meeting of the Socialist Academy, "Timiriazev as a Type of Scientist-Thinker," preserved among his personal papers held in the Russian State Archive of Social and Political History (Rossiiskii gosudarstvennyi arkhiv sotsial'no-politicheskoi istorii; hereafter, RGASPI), f. 259, op. 1, d. 44.

16. For a biography of Lunacharskii, see O'Connor 1983. For Berdiaev's recollections of his time in Vologda and meetings with Bogdanov, see Berdiaev 1949; for a recent English translation reprint, see Berdyaev 2009. On Bogdanov in Vologda, see also Novoselov 2004.

17. See Novoselov 1994, an interesting account of the Vologda exiles based on local archival materials.

18. For historical and philosophical analyses of this work, see Jensen 1978; Sadovskii 2003a, 2003b; and Andreev and Maslin 2003.

19. On the Bogdanov-Lenin rivalry and philosophical polemics, see Grille 1966; Sochor 1988; Sharapov 1997; and Lutsenko 2003.

20. For details, see Bordiugov 1995, 2; on party schools led by Bogdanov, see Rogachevskii and Michalski 1994.

21. For Bogdanov's own description of his disillusionments in political work and his former "comrades-in-arms" among Russian Marxists, see Bogdanov 1995a.

22. For an excellent English translation of the novel, see Bogdanov 1984. However, despite—and to a large extent due to—the fine literary qualities of this translation, it misses some of the nuances of the original, which are important to my arguments. Therefore, throughout the text, I use my own translations from the Russian original (Bogdanov 1908). All further quotations from the novel are from this source.

23. See, in particular, a collection of his articles entitled *New World* (Bogdanov 1905a).

24. On the comparative history of science fiction, see Rullkötter 1974; Suvin 1979; and Griffiths 1980.

25. See Suvin 1971; Ionsher 1990; Miasnikov 1999; and Shushpanov 2001.

26. Indeed, in addition to such obvious predecessors as H. G. Wells's *War of the Worlds* (1898), there were several other novels published in Russian that dealt with Mars and Martians—for instance, Vladislav Umanskii's *Unknown World: Mars and its Inhabitants* (Umanskii 1897). For an illuminating analysis of Bogdanov's novel in the contexts of both European science and science fiction, see Adams 1989.

27. On the influence of Bogdanov's Marxism on his fiction, see Jensen 1982; Stites 1984; and Graham 1984a, 1984b.

28. Similarly, H. G. Wells's choice of title for his *Modern Utopia* (1905) did not express the full meanings of the book. It is possible that Bogdanov's choice of subtitle was influenced by Wells's utopia: in a way, Bogdanov's novel presents a clear alternative to (if not a complete refutation of) the highly stratified technocratic society depicted by Wells. On the permutations of the genre of SF, see Suvin 1979. On the interrelations and differences among utopia, fantasy, and science fiction specifically in the context of postrevolutionary Russia, see Ionsher 1990.

29. On the history of blood groups, see Farr 1979; Gottlieb 1998; and Schneider 1983, 1996.

30. For a detailed analysis of Russian biologists' attitude toward this issue, see Todes 1989.

31. For a suggestive comparison of Burroughs's and Bogdanov's Mars, see Adams 1989, 8.

32. Bogdanov could have read the work in either its first English edition (Kropotkin 1902) or the first Russian edition (Kropotkin 1904). For penetrating analyses of Kropotkin's evolutionary ideas, see Todes 1989 and particularly Girón 2003. Alas, the question of Kropotkin's and, more generally, anarchist influence on Bogdanov's thinking is an almost completely neglected area in the Bogdanov scholarship (Williams 1980).

33. Indeed, a large portion of Bogdanov's 1904 article on "the gathering of man" is a direct refutation of Kropotkin's idea of cooperation as a *universal* biological trait; see Bogdanov 1904c.

34. Characteristically, in the 1913 prequel to *Red Star*, which provided the outline of Martian history prior to the socialist revolution, none of these new psychological and biological features is present, nor is there any mention of blood exchanges (Bogdanov 1913b).

35. On pre-Mendelian notions of heredity, see Müller-Wille and Rheinberger 2007.

36. On Timiriazev's views on heredity, see Gaissinovitch 1985.

37. Most likely, Bogdanov learned of the Bütschli-Maupas research through Alfred Binet, an influential French psychologist and psychiatrist, who had published several popular accounts of these experiments in the oracle of the adherents of the monistic approach to nature, *The Monist* (Binet 1890–91).

38. On the history of these experiments and their relation to Weismann's ideas, see Lustig 2000; on Weismann's work, see Winther 2001.

39. See RGASPI, f. 259, op. 1, d. 21, ll. 1–7. All the following quotations are from this source.

40. For a detailed analysis of this work, see Jensen 1978; for a recent discussion of Bogdanov's philosophy, see Soboleva 2007.

41. In her analysis of "biological ideas in tectology," Simona Poustilnik (1998) has completely neglected to investigate the sources of Bogdanov's biological ideas.

42. See Bogdanov's personnel file in ARAN, f. 350, op. 3, d. 2, l. 112.

43. See Bogdanov's letter to Lunacharskii of November 19, 1917 (Bogdanov 1990,352–55). On Narkompros and Lunacharskii, see Fitzpatrick 1970.

44. On Bogdanov and Proletkult, see Gorzka 1980; Biggart 1989; and Mally 1990.

45. It was also included in a pamphlet entitled "The Socialism of Science" (Bogdanov 1918f). For an analysis of Bogdanov's ideas of "proletarian science," see Mänicke-Gyöngyösi 1982.
46. On the Communist Academy, see David-Fox 1997; and Kozlov and Savina 2008.
47. Bogdanov 1918a, 1918b, 1918c, 1918d.
48. See Bogdanov's personnel file in ARAN, f. 350, op. 3, d. 2, ll. 104–12; here, l. 104.
49. He also published new and expanded editions of several old theoretical works (see Bogdanov 1918g, 1920a, 1920b) as well as a number of new articles and pamphlets. For a complete list of Bogdanov's publications , see Biggart, Gloveli, and Yassour 1998.

CHAPTER 4

1. For the complete text of the resolution, see *Politbiuro* . . . 2000, 1:133–34.
2. For details on the preparation of the trial, see Krasil'nikov, Morozov, and Chubykin 2002, particularly "the resolution of the plenary meeting of the Central Committee of the Russian Communist Party (the Bolsheviks) 'On Esers and Mensheviks,'" 147.
3. Documents preserved in the party archives demonstrate that the secret police had carefully monitored the Proletkult activities. See, for instance, RGASPI, f. 17, op. 84, d. 269, ll. 3–13.
4. For a biography of Krasin in English, see O'Connor 1992; on Krasin's work in Lenin's government based on previously unknown materials from Russian archives, see Khromov 2001.
5. Bogdanov's notebooks contain copious notes taken from John Maynard Keynes's book, *The Economic Consequences of the Peace* (Keynes 1920); see RGASPI, f. 259, op. 1, d. 26. This file contains several undated and unnumbered notebooks, each with its own pagination.
6. See Sobolev's reminiscences about his involvement with the circle (Sobolev 1992).
7. A copy of this paper has survived in the family archive of Bogdanov's friend Vladimir Bazarov and was recently published (Bogdanov 2003b).
8. Based on materials from the Bogdanov family archive, his biographer estimates that Bogdanov spent nearly ten thousand rubles on his experiments of 1922–26. This large sum came from royalties for Bogdanov's various books published during the same period (Iagodinskii 2006b, 194).
9. See his recollections of the arrest and imprisonment recorded after his release from prison (Bogdanov 1995b).
10. Obviously, in the eyes of the party ideologues, Bogdanov still constituted a considerable threat. One the eve of Bogdanov's arrest, one of the party-run publishers reissued the critiques of Bogdanov's philosophical views by the party's leading authorities, Vladimir Lenin and Georgii Plekhanov, under the telling title *Against Bogdanov* (Lenin and Plekhanov 1923). Two years later, another publisher issued a special volume of Plekhanov's works under the title *Against Bogdanovism* (Plekhanov 1925).
11. My reconstruction of Bogdanov's early experiments is based on his book *Struggle for Viability* (Bogdanov 1927a). Although Bogdanov used letters (X,

Y, Z) in describing the participants of the blood exchanges, most of them are easily identifiable by age, sex, and occupation provided in the descriptions. Douglas W. Huestis has recently published an English translation of the book (Bogdanov 2002b).

12. See Bogdanov's letter to the Communist Academy Presidium explaining his absence at this meeting, Bogdanov to D. B. Riazanov, February 11, 1924, in RGASPI, f. 259, op. 1, d. 63, ll. 14–14 rev.

13. Leonid Krasin to Liubov' Krasina, October 13, 1925, in Krasin 2002. In the publication, the letter is misdated October 13, 1923.

14. Leonid Krasin to Liubov' Krasina, December 4, 1925, in Krasin 2002.

15. The following reconstruction of Krasin's illness and its treatment by a blood transfusion performed by Bogdanov is based on Krasin's letters to his wife (Krasin 2002).

16. See "Sostoianie zdorov'ia L. B. Krasina" 1925.

17. The transfusion did not "rejuvenate" Krasin; he died in November 1926. See Rakovskii 1926.

18. See GARF, f. A-482, op. 42, d. 590.

19. Both Bogdanov's biographer Iagodinskii and his U.S. translator Huestis advanced this idea (Iagodinskii 2006b, 190; Huestis 2002, 6).

20. See RGASPI, f. 17, op. 84, d. 704, l. 27.

21. On the activities of the board, see *Kremlevskaia meditsina* 1997.

22. On the Central Control Commission and its functions, see Ikonnikov 1971; Rees 1987; and Getty 1997.

23. See RGASPI, f. 17, op. 84, d. 704, ll. 26–27.

24. See GARF, f. 5446, op. 31, d. 318, l. 7.

25. For the full text of the resolution, see RGASPI, f. 17, op. 2, d. 112, l. 2.

26. RGASPI, f. 17, op. 3, d. 444, l. 20. The party archives did not preserve the final draft, nor there are any records of Politburo approval of whatever Semashko and Radus-Zen'kovich had devised.

27. For a brief biography of Frunze in English, see Jacobs 1970; for a much longer, but considerably fictionalized, biography in Russian, see Arkhangel'skii 1970.

28. See "Sostoianie zdorov'ia . . ." 1925; and "Bolezn' . . ." 1925.

29. "Protokol vskrytiia" 1925; and "K istorii bolezni tov. Frunze" 1925.

30. Wild rumors were flying through the country, alleging that Frunze had been murdered on Stalin's orders. These rumors provided a plot for Boris Pilniak's famous novella *The Tale of the Unextinguished Moon*, written in January 1926 and published in May of that year in a popular literary magazine, *New World*. The censorship bureau confiscated the entire run of the magazine because of Pilniak's novella.

31. For the full text of the report, see GARF, f. 5446, op. 31, d. 318.

32. See *Politbiuro . . .* 2000, 1:421; and RGASPI, f. 17, op. 163, d. 518, l. 26.

33. For the full text of the letter, see GARF, f. 5446, op. 31, d. 318, ll. 15–17. All the following quotations are from this source.

34. RGASPI, f. 17, op. 3, d. 533, l. 10.

35. RGASPI, f. 17, op. 84, d. 704, ll. 73–95.

36. See Chernobaev 2008, 22–23. In his book, Iagodinskii provided an apocryphal account of the meeting based on "recollections" of Bogdanov's illegitimate son, Alexander Malinovskii (Iagodinskii 2006b, 190). The entire story,

139

however, seems implausible, since Malinovskii Jr. was sixteen at the time of the meeting and did not live with his father, who had quite limited contacts with his son.

37. For the complete text of the draft, see GARF, f. 5446, op. 31, d. 318, ll. 20–22.

38. See *Politbiuro* . . . 2000,1:438; and RGASPI, f. 17, op. 3, d. 547, l. 9.

39. See RGASPI, f. 17, op. 3, d. 547, l. 23. The schedule was as follows:

Thursday, 18.II: Dzerzhinskii, Menzhinskii

Friday, 19.II: Kuibyshev, Lunacharskii, Lashevich, Sadovskii, Tsiurupa, Ianson

Saturday, 20.II: Kamenev, Unshlikht, Zinov'ev, Rudzutak, K. Tsetkin, Gorbunov

Monday, 22.II: Sol'ts, Iagoda, Iaroslavskii, Litvinov, Sosnovskii, Tomskii, Solov'ev

Tuesday, 23.II: Trotskii, Briukhanov, Chicherin, Ol'minskii

Wednesday, 24.II: Shliapnikov, Liadov, Bukharin, Pokrovskii, Uglanov, Krupskaia

Thursday, 25.II: Voroshilov, Smidovich, Leplevskii, Shvedchikov, Kiselev

Friday, 26.II: Taratuta, Molotov, Ter-Gabrielian, Stalin, Rykov

40. For official information on the visit, see "Chestvovanie professorov . . ." 1926. On secret arrangements for the consultation, see GARF, f. 5446, op. 31, d. 318, l. 127; and RGASPI, f. 17, op. 3, d. 547, l. 23.

41. For details on the intricate structure of the Soviet government apparatus and relations among its numerous agencies, see Korzhikhina 1994.

42. See GARF, f. A-259, op. 10a, d. 5, l. 92.

43. Today it houses the French embassy.

44. On Bogdanov's dispute with the union, see GARF, f. A-259, op. 12a, d. 47, ll. 97–98; see also "V Sovnarkome RSFSR" 1926.

45. I was unable to find this document in Narkomzdrav's archives, but Bogdanov cited it extensively in his brochure (Bogdanov 1927b, 4). A slightly revised version of the brochure was included in the first volume of the Bogdanov institute's *Proceedings* (Bogdanov, Bogomolets, and Konchalovskii 1928); Huestis translated this version in its entirety into English (Bogdanov 2002a, 217–70).

46. On the history of the laboratory to study Lenin's brain, see Richter 2007; on the history of the laboratory for racial pathology, see Weindling 1992.

47. *MMZh*, 1926, 5:130.

48. All these cases are described in Bogdanov 1927b, 6.

49. Although the title page of the book gave 1927 as the year of publication, the book was actually released in December 1926 (see Ottenberg 1926).

50. A brief account of his theory also appeared in the *Russo-German Medical Journal* (Bogdanov 1927d).

CHAPTER 5

1. For dated but still useful general overviews of the studies of senescence, see Trimmer 1967 and Gruman 1966; the latter work has recently been

reissued (Gruman 2003). For a very thorough bibliographic survey of literature (mostly in English), see Freeman 1979. For more recent and more popular works, see Benecke 2002; Hall 2003; and Boia 2004.

2. On Claude Bernard and his concept of internal secretions, see Coleman 1985; and Wasserstein 1996.

3. On Brown-Sequard and the rise of organotherapy, see Borell 1976a, 1976b, 1985. On the development of organotherapy and endocrinology in Russia, see Krementsov 2008.

4. Unfortunately, there is no book-length examination of Metchnikoff's ideas of prolongevity. For a brief discussion, see Cochrane 2003; Aveline 2003; and Hugh 2003.

5. For overviews of the history of gerontology in the U.S. context, see Achenbaum 1995; and Katz 1996.

6. For a thorough contemporary overview of this concept, see Korschelt 1922. Bogdanov used a 1925 Russian edition of this book in his analysis of senescence. For a similar survey in English, see Robertson 1923.

7. On Steinach's works, see Sengoopta 2006; on Voronoff's works, see Hamilton 1986. Unfortunately, neither book even mentions the situation in Russia.

8. On "Vivarium" and its remarkable cohort of researchers, see Reiter 1999.

9. See Palladin 1923; Shmidt 1923, 1924; Shtark 1923; London and Kryzhanovskii 1924; Gremiatskii 1925; and many others.

10. *Omolozhenie v Rossii* 1924.

11. See, for instance, Damskii 1923; Rodionov 1924; Kogan 1924; Sukhov 1925; and many others.

12. See, for instance, Fabrikant and Taft 1923; Rodionov 1923; Fridman 1923; Shustrov, Karpova, and Tikhomirov 1924; Shustrov 1924; and Kustria 1924.

13. See report on the congress in *Turkestanskii meditsinskii zhurnal* 3, no. 9 (1924): 223–26.

14. See "2-i nauchnyi s"ezd . . ." 1925; "Vtoroi nauchnyi s"ezd . . ." 1925; Fedorovich and Kaizer 1925; "Oblastnoi s"ezd . . ." 1925; and "Kratkii otchet . . ." 1925.

15. See *Trudy VII Vsesoiuznogo s"ezda terapevtov* 1926,523–30.

16. For a contemporary overview, see Shirokogorov 1924.

17. See "Ne omolozhenie, a osvezhenie" 1924.

18. For instance, the Georgian surgeon N. Dzhaparidze reported that out of nine cases of Steinach's operation he had performed at his clinic, seven produced no effect, and two improved the conditions of his patients "insignificantly" ("Kratkii otchet . . ." 1925, 444–46).

19. See "Disput ob omolozhenii" 1924.

20. See "Teoriia prof. Shteinakha v kino" 1922; and "Kino. Beseda s professorom Kol'tsovym . . ." 1923.

21. See *Pravda*, June 4, 1924, 7.

22. See "Kuda ustraivat' ekskursii" 1924.

23. See "Kino, Kul'tob"edinenie" 1924.

24. See Voskresenskii 1924; and "Ne omolozhenie, a osvezhenie" 1924.

25. "Kogo nado omolazhivat'" 1925; "Kogo nuzhno omolazhivat'" 1925.

26. See Anichkov 1920; Isaev 1922; Nemilov 1923a, 1923b, 1925; Ryzhkov 1923; "Omolozhenie" 1923b; Azimov 1923; Vasilevskii 1925a, 1925b; and many others.

27. "Opyty professora Voronova" 1922; "O tak-nazyvaemom omolozhenii" 1923; "Kino. Beseda s professorom Kol'tsovym . . ." 1923; "Omolozhenie" 1923a; "V dome uchenykh" 1923; "Omolozhenie liudei i zhivotnykh" 1923; "Opyt omolozheniia" 1923; "Novosti nauki i tekhniki . . ." 1924; and Kol'tsov 1923c.

28. See "Blestiashchie opyty omolozheniia" 1922; S. V. 1922; Gri-v 1923; and many others.

29. See "Disput ob omolozhenii" 1924; "Ne omolozhenie, a osvezhenie" 1924; "Omolozhenie liudei (beseda s prof. Zavadovskim)" 1924; "D-r S. Voronov ob omolozhenii liudei" 1924; and "Samoomolozhenie d-ra Voronova" 1924.

30. "Omolozhenie Lloid-Dzhorzha" 1924; and "1500 omolodivshikhsia" 1924.

31. See Bubus 1925; "Omolozhenie zhivotnykh . . ." 1925; "Posle omolozheniia . . ." 1925; "Bor'ba protiv omolozheniia . . ." 1925; "Nauka vyshe boga" 1925; "Kogo nado omolazhivat'" 1925; "Pokhod protiv omolozheniia" 1925; Vasilevskii 1925a; "Kogo nuzhno omolazhivat'" 1925; "D-r Voronov sobiraetsia v Moskvu" 1925; and "Pervye itogi rabot prof. Voronova" 1925.

32. See, for example, Howell 2006.

33. On the history of sex hormones, see Oudsboorn 1994; and Sengoopta 2006.

34. Today, Jaworski is mostly remembered as an originator of the idea of "living planet" (Twentyman 1971).

35. Either Bogdanov himself or someone in his circle probably read *Evening Moscow* on a regular basis. In his *Izvestiia* article, Bogdanov (1926) directly referred to Bruskin's interview published by the newspaper a few months earlier. Furthermore, beginning in April 1926, the newspaper regularly carried articles on the new institute of blood transfusion, including two interviews with its director, Bogdanov ("Otkrytie instituta . . ." 1926; "Institut perelivaniia krovi . . ." 1926; "Chem budet zanimat'sia . . ." 1926; and "V institute perelivaniia krovi" 1926).

36. For a detailed analysis of this polemical style and its "infusion" into scientific debates in the 1920s and 1930s, see Krementsov 1997, 25–30, 45–53.

37. On Kammerer's life story, see Koestler 1971; on Kammerer's research, see Gliboff 2006.

38. Compare Bogdanov's dismissive critique with a detailed four-page-long review of Bruskin's work by a well-known professor of surgery, Vladimir Pokotilo (1927).

CHAPTER 6

1. This short review was perhaps written by Nikolai Nikolaevich Elanskii.

2. For a rave review of the brochure, as well as Bogdanov's *Struggle for Viability*, which appeared in *Evening Moscow*, see Poltavskii 1928.

3. For a biography of Bogomolets, see Sirotinin 1967.

4. See Vlados and Shustrov 1927. It is also possible that there was another reason Bogdanov invited Vlados to his institute. In 1924, Vlados published an article on "auto-hemo-therapy," the stimulation of blood production in anemic patients by the intramuscular injections of their own blood, which resonated well with Bogdanov's theory (Vlados and Tareev 1924).

5. Bogdanov recounted this and other mishaps in a letter of resignation he drafted in January 1928. For the full text of the letter, see RGASPI, f. 259, op. 1, d. 47, ll. 1–12.

6. For the brochure and a protocol of the meeting, see GARF, f. A-482, op. 1, d. 635, ll. 288–94.
7. On possible causes of Bogdanov's death from the viewpoint of modern medicine, see Huestis 1996.
8. See "Pamiati A. A. Bogdanova" 1928a; and "U groba A. A. Bogdanova" 1928.
9. On the organization of Bogdanov's funeral, see "Komissiia . . ." 1928; "Pokhorony A. A. Bogdanova" 1928; and "Pokhorony A. A. Bogdanova (Malinovskogo)" 1928.
10. On the commemoration of Bogdanov, see "Pamiati A. A. Bogdanova" 1928a, 1928b, 1928c; also GARF, f. A-259, op. 12a, d. 45, l. 13. On naming the institute after Bogdanov, see "Pereimenovanie Instituta . . ." 1928; also GARF, f. A-482, op. 1, d. 622, ll. 146–47; and f. A-259, op. 13, d. 111, l. 10.
11. See Bogdanov 1929a and 1929b.
12. For details of Vasilevskii's participation in a blood exchange and its aftermath, see Bogdanov's letter to Vasilevskii, February 21, 1928, in the Russian State Archive of Literature and the Arts, f. 108, op. 1, d. 4.
13. See also Vasilevskii's publications in another popular journal, which followed exactly the same pattern (Vasilevskii 1928c, 1928e).
14. The plan is preserved in GARF, f. A-482, op. 1, d. 640, ll. 406–406 rev.; d. 639, ll. 59–60.
15. For an analysis of Bogomolets's views on blood transfusions, see Koziner 1987.
16. The only time Bogdanov's ideas were mentioned in passing in the entire volume was in the chapter entitled "Clinical Evaluation of the Conditions for the Use of Blood Transfusion as a Therapeutic Method," written by Vlados. See also Konchalovskii 1931.
17. For instance, a long popular article on blood transfusions written by the prominent surgeon Vladimir Oppel' for the weekly magazine *Hygiene and Health of the Worker's and Peasant's Family* did not even mention Bogdanov's name (Oppel' 1930).
18. See "Predstavliaet li opasnost' perelivanie krovi?" 1928, 1; italics in the original.
19. See, for instance, a frank assessment of the state of affairs in the field of transfusions in the country delivered at the Twenty-first Congress of Russian Surgeons in Leningrad in 1929 (Barinshtein and Liubarskii 1930).
20. Accounts of this meeting appeared in both professional ("Protokol zasedaniia" 1926) and popular ("Perelivanie krovi" 1925a) periodicals. All the following quotations are from these sources.
21. The resolution was included in the publication of Gesse's report (Gesse 1926a).
22. See report on the congress in *NKhA* 10, no. 3 (1926): 422.
23. See, for instance, "O perelivanii krovi" 1923a, 1923b; "Perelivanie krovi" 1925b; and "Prodazhnaia tsena krovi" 1925. For an excellent cultural history of blood transfusions in the United States, see Lederer 2008.
24. Obviously, Dembo did not read a barbed article on "Selling women's milk" in the United States, published a year earlier by a popular Leningrad daily, *Red Gazette* (see "Torgovlia zhenskim molokom" 1925).
25. For an account of the institute's work with donors, see Emel'ianov 1928.
26. See, for instance, Elanskii's address "On the Question of the Organization of Blood Transfusions," delivered to the Pirogov Society of Russian Surgeons

on November 28, 1930, reported in *Physician's Gazette* (*Vrachebnaia gazeta* 4 [1931]: 324–25).

27. For a contemporary account of the development of transfusions with cadavers' blood, see Skundina 1935.

28. For an analysis of the "revolution from above," see Tucker 1990; see also Fitzpatrick 1978.

29. For details on the effects of the "revolution from above" on the system of Soviet science, see Krementsov 1997, 31–53.

30. On the militarization of Soviet Russia, see Stone 2000. Unfortunately, this excellent analysis does not even mention the Red Army's medical services.

31. For a copy of the list, see ARAN, f. 570, op. 1, d. 29, ll. 101–3.

32. These dramatic developments were recorded in a jubilee brochure commemorating the tenth anniversary of the Bogdanov Institute (*X let . . .* n.d.).

33. On the importance of such systems in other countries, see Schneider 1997.

34. On the "volunteer" societies in the USSR, see Odom 1973.

35. For a complete text of the decree, see "'O kadrakh donorov'" 1935.

CHAPTER 7

1. For the Russian original, see Bogdanov 1924b,167–69. For an English translation, see Bogdanov 1984, 235–39.

2. See Bogdanov's notebooks in RGASPI, f. 259, op. 1, d. 26.

3. For the complete text of the letter, see RGASPI, f. 259, op. 1. d. 84, ll. 1–6.

4. See, for example, Williams 1980; on the long history of the debate in Russia, see Khakhordin 1999.

5. On the "purification" of blood, see Chepov 1930; on the history of Soviet research on "hormone supplements," see Naiman 2002.

6. This is further demonstrated by the fact that Bogomolets himself continued research on his cytotoxic "rejuvenation serum" well into the 1940s, until his own death; for a sample of his English-language publications, see Bogomolets 1945a, 1945b.

7. Bogdanov was not unique but rather typical in this respect. Indeed, this "linguistic" particularity had significant implications for Russian science and culture more generally. The distinction and division between the "two cultures"—natural sciences and the humanities—quite prominent in the English-speaking countries, simply did not exist in late nineteenth- and early twentieth-century Russia. Many institutions—from the Imperial (and later Soviet) Academy of Sciences to the Psycho-Neurological Academy—united under one roof natural and social sciences and the humanities. Many Russian scientists dabbled in the humanities, publishing fiction and poetry, writing music, and patronizing the arts, while many writers considered their own trade as a "method of cognition" equal in its cognitive powers to natural sciences. Works of fiction and results of scientific research were published side by side in the same "thick" journals, such as *The Contemporary*, *Russian Wealth*, and *Russian Thought*, to which Bogdanov regularly contributed.

8. "Investigative" practices and their impact on the "production" of knowledge have attracted considerable attention from historians and sociologists of science (e.g., Fleck 1979; Latour and Woolgar 1976; Mendelsohn, Weingart,

and Whitley 1977; Clarke and Fujimura 1992; Pickering 1992; Kohler 1994; Pickstone 2001; Todes 2002; and many others). The "social" practices that "produce" formal and informal organizations, patrons, careers, roles, and policies have attracted significantly less attention (e.g., Abir-Am 1992; Golinsky 1995; Abir-Am and Elliott 1999). For analyses of Soviet scientists' "social" practices, see Krementsov 1997, 2005, 2008.

9. RGASPI, f. 259, op. 1. d. 84, l. 5.

10. This reasoning has an interesting consequence: if two "rational" human beings disagree in the assessment of the same data, it must reflect the difference in their individual class consciousness. This logic was definitely at play in numerous intraparty debates and constituted a major element of the Soviet critical style. Every self-proclaimed "Marxist" always construed any disagreement with his or her views as the manifestation of the opponent's "bourgeois" class origins and/or ideology and accordingly immediately labeled the opponent "anti-Marxist." The argument worked in the other way too: if the opponent belonged (by birth or education) to the "bourgeoisie," "peasantry" or "intelligentsia," he or she could almost never be a "true" Marxist.

11. Bogdanov to Lunacharskii, November 19, 1917, in Bogdanov 1990,352–55. See also his letters to Dzerzhinskii in Bogdanov 1995b.

12. See GARF, f. A-482, op. 1, d. 635, ll. 288–94.

13. The two books were Pauchet and Bécart 1924; and Weil and Isch-Wall 1925.

14. See RGASPI, f. 259, op. 1, d. 26.

15. For a nearly complete bibliography of publications on blood groups and blood transfusions from 1900 to 1933, see Kenig 1936.

16. *Bogdanovshchina* (see Plekhanov 1925). In the 1930s, Lenin's *Materialism and Empiriocriticism* became the "bible" of Soviet dialectical materialism, naturally transforming Bogdanov's philosophical ideas into a favorite practice target for numerous party ideologues and demagogues (see, for example, Shcheglov 1937). On the process of the transformation of Bogdanov's philosophy into "Bogdanovism," see Joravsky 1961.

17. One can suggest that Boris Hessen's (1931) seminal work on "The Social and Economic Roots of Newton's *Principia*," which exerted profound influence on the development of the Marxist history of science, presents a similar example of the extensive application of Bogdanov's ideas about science without acknowledging their originator.

References

2-i nauchnyi s"ezd vrachei Srednei Azii, Tashkent, April 26–30, 1925. 1925. *Turkestanskii meditsinskii zhurnal* 4 (9): 561.

1500 omolodivshikhsia. *VM*, September 15, 1924, 2.

X let Tsentral'nogo instituta gematologii i perelivaniia krovi imeni A. A. Bogdanova, 1926–1936. N.d. Moscow: n.p.

Abir-Am, Pnina G., ed. 1992. Special issue on the historical ethnography of scientific rituals. *Social Epistemology* 6 (4).

Abir-Am, Pnina G., and Clark A. Elliott, eds. 1999. *Commemorative practices in science. Osiris* 14.

Adams, Mark B. 1989. *Red star*: Another look at Alexander Bogdanov. *Slavic Review* 48 (1): 1–15.

———. 1990. Soviet Union: The Russian research system. In *The academic research enterprise within the industrialized nations: Comparative perspectives*, 51–65. Washington, DC: National Academy Press.

———. 2000. Last judgment: The visionary biology of J.B.S. Haldane. *JHB* 33:457–91.

———. 2004. The quest for immortality: Visions and presentiments in science and literature. In *The fountain of youth: Cultural, scientific, and ethical perspectives on a biomedical goal*, ed. Stephen G. Post and Robert H. Binstock, 38–71. New York: Oxford University Press.

Achenbaum, W. Andrew. 1995. *Crossing frontiers: Gerontology emerges as a science*. New York and Cambridge, England: Cambridge University Press.

Afonin, N. I. 2002. Istoricheskii ocherk razvitiia transfusiologii v Rossii. *Vestnik sluzhby krovi Rossii* 2:42–51.

Alekseev, V. V. 2006. I. V. Stalin and A. A. Bogdanov: Materialy dlia issledovaniia problemy. *Vestnik Kamskogo instituta gumanitarnykh i inzhenernykh tekhnologii* 1:32–37.

Allen, Garland E. 1975. *Life science in the twentieth century*. New York: Wiley.

———. 1991. Mechanistic and dialectical materialism in 20th century evolutionary theory: The work of Ivan I. Schmalhausen. In *New perspectives on evolution*, ed. Leonard Warren and Hilary Koprowski, 15–36. New York: Wiley-Liss.

Andreev, A. L., and M. A. Maslin. 2003. A. A. Bogdanov kak filosof i sotsial'nyi myslitel. In Bogdanov 2003a, 366–76.

Anfimov, Ia. A. 1893. *Soznanie i lichnost' pri dushevnykh zabolevaniiakh*. Tomsk: P. I. Makushin.

———. 1915. *Programma po nervnym i dushevnym bolezniam na zvanie lekaria*. Khar'kov: B. Bengis.

Anichkov, N. N. 1920. O iavleniiakh starosti u cheloveka i zhivotnykh. *ChiP* 1:61–63.

Arkhangel'skii, Vl. 1970. *Frunze*. Moscow: Molodaia gvardiia.

Arutiunian, M. S. 1932. Primenenie konservirovannoi krovi. *Sovremnnye problemy perelivaniia krovi i gematologii* 3–4:38–50.

Aveline, Mark. 2003. Elie Metschnikoff and his theory of an "instinct de la vie." *International Journal of Epidemiology* 32 (1): 32–36.

Azimov, G. I. 1923. Bor'ba so starost'iu i smert'iu. *Iskra* 2:21–27.

Bailes, Kendall E. 1978. *Technology and society under Lenin and Stalin: Origins of the Soviet technical intelligentsia, 1917–1941*. Princeton, NJ: Princeton University Press.

Bakunin, N. 1927. Zhivaia voda dvadtsatogo veka. *Iskry nauki* 9:337–38.

Balakhovskii, S. D., F. G. Ginzburg, T. A. Palitsyna, S. B. Rzhekhina, and R. S. Farberova. 1932. Konservirovanie krovi, prednaznachennoi dlia transfuzii. *Sovremnnye problemy perelivaniia krovi i gematologii* 3–4:16–37.

Ballestrem, Karl G. 1969. Lenin and Bogdanov. *Studies in Soviet Thought* 9:283–310.

Banerjee, Anindita. 2000. The genesis and evolution of science fiction in fin de siècle Russia, 1880–1921. PhD diss., University of California, Los Angeles.

Baranovskii, B. 1927. Institut perelivaniia krovi. *Pravda,* July 27, 8.

Barinshtein, L. A. 1924. Perelivanie krovi, v svete sovremennykh dannykh. *Sovremennaia meditsina* 1:39–52.

———. 1928. *Perelivanie krovi (teoriia i praktika)*. Odessa: n.p.

Barinshtein, L. A., and B. I. Liubarskii. 1930. Organizatsionnaia storona voprosa o perelivanii krovi. In *XXI-i s"ezd rossiiskikh khirurgov, Leningrad, 5–9 iiunia 1929*, 131–39. Moscow: Obshchestvo rossiiskikh khirurgov.

Barskii, Kh. G. 1927. Izoaglliutinatsionnaia kharakteristika chelovecheskoi krovi i perelivanie krovi. *Trudy 7-go Vsesoiuznogo s"ezda ginekologov i akusherov*, 816–18. Leningrad: Prakticheskaia meditsina.

Basile, G. M. 1984. The utopia of rebirth: Aleksandr Bogdanov's "Krasnaia zvesda." *Canadian American Slavic Studies* 18 (1–2): 54–62.

Bastrakova, M. S. 1973. *Stanovlenie sovetskoi sistemy organizatsii nauki, 1917–1922*. Moscow: Nauka.

Bauduin, Philippe. 2007. *L'or rouge: Les alliés et la transfusion sanguine, Normandie 44*. Le Coudray-Macouard: Cheminements.

Belen'kii, D. N. 1930. O konservirovannoi krovi. *NKhA* 22 (1): 25–47.

Belitskaia, A. Ia. 1974. Aleksandr Aleksandrovich Bogdanov i ego zhiznennyi podvig. *Sovetskoe zdravookhranenie* 3:73–77.

Belova, A. A. 1974. *A. A. Bogdanov*. Moscow: Meditsina.

———. 1983. *Istoriia stanovleniia i razvitiia ucheniia o mekhanizme deistviia perelivaniia krovi*. Tbilisi: Sabchota Sakartvelo.

Benecke, M. 2002. *The dream of eternal life: Biomedicine, aging, and immortality*. New York: Columbia University Press.

Berdiaev, N. 1949. *Samopoznanie (opyt filosofskoi biografii)*. Paris: YMCA-Press.

Berdyaev, Nicolas. 2009. *Self-knowledge: An essay in autobiography*. San Rafael, CA: Semantron Press.

Biggart, John. 1989. Alexander Bogdanov, left-bolshevism and the Proletkult, 1904–1932. PhD diss., University of East Anglia, Norwich.

Biggart, John, Peter Dudley, and Francis King, eds. 1998 *Alexander Bogdanov and the origins of systems thinking in Russia.* Aldershot, England, and Brookfield, VT: Ashgate.

Biggart, John, Georgii Gloveli, and Avraham Yassour, eds. 1998. *Bogdanov and his work: A guide to the published and unpublished works of Alexander A. Bogdanov (Malinovsky), 1873-1928.* Aldershot, England: Ashgate.

Billington, James H. 1958. *Mikhailovsky and Russian populism.* Oxford, England: Clarendon Press.

Binet, Alfred. 1890-91. The immortality of infusoria. *The Monist* 1:20-37.

Biuleteni postiinoi komisii vivchania krov'ianikh ugrupovani, or *Verhandlungen der stadigen commission fur blutgruppenforschung (Ukrainisches Zentralblatt fur Blutgruppenforschung).* 1927. Bd. 1, H. 1.

Blestiashchie opyty omolozheniia. 1922. *Poslednie novosti,* July 3, 1.

Blinkov, N. 1925. Perelivanie krovi. *ChiP* 12:17-24.

Bogdanov, A. 1897. *Kratkii kurs ekonomicheskoi nauki.* Moscow: A. M. Murinova.

———. 1899. *Osnovnye elementy istoricheskogo vzgliada na priorodu: Priroda. Zhizn'. Psikhika. Obshchestvo.* St. Petersburg: Seiatel'.

———. 1902. Razvitie zhizni v prirode i obshchestve. *Obrazovanie* 4:33-46; 5-6:91-110; 7-8:66-83.

———. 1903a. Ideal poznaniia. *Voprosy filosofii i psikhologii* 67:186-233.

———. 1903b. Zhizn' i psikhika. *Voprosy filosofii i psikhologii* 69:682-708; 70:824-66.

———. 1904a. *Iz psikhologii obshchestva (stat'i 1901-1904 gg).* St. Petersburg: S. Dorovatovskii i A. Charushnikov.

———. 1904b. Psikhicheskii podbor. *Voprosy filosofii i psikhologii* 73:335-79; 74:485-519.

———. 1904c. Sobiranie cheloveka. *Pravda* (Moscow) 4:158-75.

———. 1904-6. *Empiriomonizm: Stat'i po filosofii.* 3 vols. Moscow: S. Dorovatovskii i A. Charushnikov.

———. 1905a. *Novyi mir (Stat'i 1904-1905).* Moscow: S. Dorovatovskii i A. Charushnikov.

———. 1905b. Tseli i normy zhizni. *Obrazovanie* 7:265-98.

———. 1908. *Krasnaia zvezda.* St. Petersburg: T-vo khudozhnikov pechati.

———. 1909a. Filosofiia sovremennogo estestvoispytatelia. In *Ocherki filosofii kollektivizma,* 35-142. St. Petersburg: Znanie.

——— [N. Verner]. 1909b. Nauka i filosofiia. In *Ocherki filosofii kollektivizma,* 9-34. St. Petersburg: Znanie.

———. 1910. *Padenie velikogo fetishizma (sovremennyi krizis ideologii): Vera i nauka (o knige V. Il'ina, Materializm i empiriokrititsizm).* Moscow: S. Dorovatovskii i A. Charushnikov.

———. 1913a. *Filosofiia zhivogo opyta.* Petrograd: M. I. Semenov.

———. 1913b. *Inzhener Menni (fantasticheskii roman).* Moscow: S. Dorovatovskii i A. Charushnikov.

———. 1913c. Taina nauki. *Sovremennik* 8:161-82.

———. 1913d. *Vseobshchaia organizatsionnaia nauka (tektologiia).* St. Petersburg: M. I. Semenov.

———. 1917a. *Vseobshchaia organizatsionnaia nauka (tektologiia).* Vol. 2. Moscow: Knizhnoe izdatel'stvo pisatelei.

————. 1917b. *Zadachi rabochikh v revolutsii*. 3rd ed. Moscow: Sazonov.

————. 1918a. *Inzhener Menni: Fantasticheskii roman*. Petrograd: Petrogradskii sovet rabochikh i krasnoarmeiskikh deputatov.

————. 1918b. *Inzhener Menni: Fantasticheskii roman*. Moscow: Volna.

————. 1918c. *Krasnaia zvezda: Roman-utopia*. Petrograd: Petrogradskii sovet rabochikh i krasnoarmeiskikh deputatov.

————. 1918d. *Krasnaia zvezda: Roman-utopia*. Moscow: Knigoizdatel'stvo pisatelei.

————. 1918e. *Nauka i rabochii klass*. Moscow: n.p.

————. 1918f. *Sotsializm nauki*. Moscow: Proletarskaia kul'tura.

————. 1918g. *Voprosy sotsializma*. Moscow: Knigoizdatel'stvo pisatelei.

————. 1920a. *Filosofiia zhivogo opyta*. Moscow: GIZ.

————. 1920b. *Novyi mir*. Moscow: GIZ.

————. 1920c. Ocherki organizatsionnoi nauki. *Proletarskaia kul'tura* 15–16:6–38.

————. 1920d. Pamiati K. A. Timiriazeva. *Proletarskaia kul'tura* 15–16:1–3.

————. 1922. *Tektologiia*. Berlin: Grzhebin.

————. 1924a. *Krasnaia zvezda*. Leningrad and Moscow: Kniga.

————. 1924b. Marsianin, zabroshennyi na zemliu. In Bogdanov 1924a, 167–69.

————. 1926. Institut perelivaniia krovi. *Izvestiia*, April 4, 5.

————. 1927a. *Bor'ba za zhiznesposobnost'*. Moscow: Novaia Moskva.

————. 1927b. *God raboty instituta perelivaniia krovi, 1926–27*. Moscow: Institut perelivaniia krovi.

————. 1927c. Malen'koe nachalo bol'shogo dela. *Izvestiia*, January 21, 7.

———— [Bogdanow]. 1927d. Zur Theorie des Alterns. *Russko-nemetskii meditsinskii zhurnal* 3 (1): 32–44.

————. 1929a. *Inzhener Menni: Fantasticheskii roman*. Leningrad: Krasnaia gazeta.

————. 1929b. *Krasnaia zvezda: Roman-utopiia*. Leningrad: Krasnaia gazeta.

————. 1977. *La science, l'art et la classe ouvrière*. Translated from the Russian and annotated by Blanche Grinbaum. Presentations of Henri Deluy and Dominique Lecourt. Paris: F. Maspero.

————. 1984. *"Red star": The first Bolshevik utopia*. Ed. Loren R. Graham and Richard Stites. Trans. Charles Rouge. Bloomington: Indiana University Press.

————. 1989. *Tektologiia: Vseobshchaia organizatsionnaia nauka*. 2 vols. Moscow: Ekonomika.

————. 1990. *Voprosy sotsializma: Raboty raznykh let*. Moscow: Politizdat.

————. 1995a. Desiatiletie otlucheniia ot marksizma: Iubileinyi sbornik (1904–1914). In Bordiugov 1995, 3:1–132.

————. 1995b. Piat' nedel' v GPU (8 sentiabria–13 octiabria 1923 goda). In Bordiugov 1995, 1:34–44.

————. 2002a. The first year's work of the Institute of Blood Transfusion. In Bogdanov 2002b, 217–70.

————. 2002b. *The struggle for viability: Collectivism through blood exchange*. Ed. and transl. Douglas W. Huestis. Philadelphia: Xlibris.

————. 2003a. *Empiriomonizm*. Moscow: Respublika.

————. 2003b. O fiziologicheskom kollektivizme. *VMIAB* 14 (2). http://www.bogdinst.ru/vestnik/v14_05.htm (accessed September 12, 2004).

Bogdanov, A., A. A. Bogomolets, and M. P. Konchalovskii, eds. 1928. *Na novom pole: Trudy Gosudarstvennogo nauchnogo instituta perelivaniia krovi imeni A. A. Bogdanova*. Vol. 1. Moscow: Izdatel'stvo Instituta perelivaniia krovi.

Bogomolets, A. A. 1928. Pamiati A. A. Bogdanova: K voprosu o nauchnom i prakticheskom znachenii metoda perelivaniia krovi. In Bogdanov, Bogomolets, and Konchalovskii 1928, XXVII–XXXI.

———. 1929a. Nauchnoe i prakticheskoe znachenie metoda perelivaniia krovi. *Nauchnoe slovo* 8:35–64.

———. 1929b. Novoe o perelivanii krovi. *VM*, March 21, 3.

———. 1945a. Anti-reticular cytotoxic serum as a means of pathogenic therapy. *Canadian Medical Association Journal* 52:76–77.

———. 1945b. Blood transfusion in treatment of internal disease. *American Review of Soviet Medicine* 2:196–98.

Bogomolets, A. A., M. P. Konchalovskii, and S. I. Spasokukotskii, eds. 1930. *Perelivanie krovi kak lechebnyi metod: Pokazaniia, tekhnika transfuzii i trebovaniia, pred"iavliaemye k donoru*. Moscow: Gosmedizdat.

Boia, Lucian. 2004. *Forever young: A cultural history of longevity*. London: Reaktion Books.

Bolezn' narkomvoenmora M. V. Frunze. 1925. *Pravda*, October 31, 2.

Boothby W. M., and V. N. Shamoff [Shamov]. 1915. A study of the late effect of division of the pulmonary branches of the vagus nerve on the gaseous metabolism, gas exchange, and respiratory mechanisms in dogs. *American Journal of Physiology* 37:418–30.

Bor'ba protiv omolozheniia v Indii. 1925. *VM*, June 12, 3.

Bordiugov, G. A., ed. 1995. *Neizvestnyi Bogdanov*. 3 vols. Moscow: AIRO-XX.

Borell, Merriley. 1976a. Brown-Séquard's organotherapy and its appearance in America at the end of the 19th century. *Bulletin of the History of Medicine* 50:309–20.

———. 1976b. Organotherapy, British physiology, and discovery of the internal secretions. *JHB* 9:235–68.

———. 1985. Organotherapy and the emergence of reproductive endocrinology. *JHB* 18:1–30.

Braitsev, V. Ia. 1932. Problema transporta konservirovannoi krovi. *NKhA* 27 (1): 53–64.

Breitman, M. Ia. 1927. A. Bogdanov, Bor'ba za zhiznesposobnost'. *Vrachebnaia gazeta* 15:1142.

Brikner, F., F. Suponitskaia, and A. Charnyi. 1926. Eksperimental'naia otsenka vlivaemykh pri bol'shikh krovopoteriakh rastvorov. *ZhDUV* 5:285–98.

Britikov, A. F. 1970. *Russkii sovetskii nauchno-fantasticheskii roman*. Leningrad: Nauka.

Bruskin, Ia. M. 1924. Sovremennaia germanskaia khirurgiia. *NKhA* 5 (17–20): 162–95.

———. 1925a. Perelivanie krovi. *Meditsinskii rabotnik* 18:13–14.

———. 1925b. Perelivanie krovi i noveishie uspekhi v etoi oblasti. *VSM* 1:7–10; 2:8–11.

———. 1926. Perelivanie krovi. *Molodaia gvardiia* 1:208–12.

———. 1927. *Perelivanie krovi*. Moscow: NKZdrav.

Bubus. 1925. Obez'iany i chelovechestvo. *VM*, February 23, 3.

Bukharin, N. 1921a. Kollektivisticheskoe likvidatorstvo. *Pravda*, December 13, 1.

———. 1921b. K s"ezdu Proletkul'ta. *Pravda*, November 21, 1.

———. 1928. A. A. Bogdanov. *Pravda*, April 8, 5.

———. 1932. *Darvinizm i marksizm*. Leningrad: Akademiia nauk.

Burlachenko, A. G., and K. Gess-de-Kal've. 1925. K voprosu ob izmenenii krovi (morfologii i fermenta lipazy) posle autotransfuzii gemolizirovannoi krovi. *Trudy pervogo oblastnogo s"ezda khirurgov levoberezhnoi Ukrainy*, 1–5. Kharkov: Nauchnaia assotsiatsiia Medsantrud.

Carrel, A. 1907. The surgery of blood vessels. *Bulletin of the Johns Hopkins Hospital* 18:18–28.

Chem budet zanimat'sia institut perelivaniia krovi. 1926. *VM*, July 19, 2.

Chepov, P. M. 1930. O prizhiznennom promyvanii krovi. *ChiP* 22:17–22.

Chernobaev, A., ed. 2008. *Na prieme u Stalina*. Moscow: Novyi Khronograf.

Chestvovanie professorov F. Krauza i O. Ferstera. 1926. *Izvestiia*, February 28, 5.

Chistova, S. P. 1967. Rol' A. A. Bogdanova v razvitii Sovetskoi meditsiny. *Sovetskaia meditsina* 30 (6): 147–50.

Clarke, Adele E., and Joan H. Fujimura, eds. 1992. *The right tools for the job: At work in twentieth century life sciences*. Princeton, NJ: Princeton University Press.

Cochrane, A. L. 2003. Elie Metschnikoff and his theory of an "instinct de la mort." *International journal of epidemiology* 32 (1): 32–36.

Coleman, William. 1985. The cognitive basis of the discipline: Claude Bernard on physiology. *Isis* 76:49–70.

Crile, G. W. 1906a. Direct transfusion of blood in the treatment of hemorrhage: Preliminary clinical note. *JAMA* 47:1482–84.

———. 1906b. Experimental and clinical observations upon direct transfusion of blood. *Science* 24:765.

———. 1907. The technique of direct transfusion of blood. *Annals of Surgery* 46:329–32.

———. 1909. *Hemorrhage and transfusion*. New York and London: D. Appleton & Co.

———. 1947. *An autobiography*. 2 vols. Ed. Grace Crile. Philadelphia: Lippincott.

Damskii, A. Ia. 1923. K probleme omolozheniia. *Nauchnye izvestiia gosudarstvennogo smolenskogo universiteta* 1:119–26.

David-Fox, Michael. 1997. *Revolution of the mind: Higher learning among the Bolsheviks, 1918–1929*. Ithaca, NY: Cornell University Press.

Dembo, L. I. 1926. Iuridicheskie predposylki k voprosu o platnom donorstve. *Leningradskii meditsinskii zhurnal* 6:62–65.

Diamond, Louis K. 1980. A history of blood transfusion. In *Blood, Pure and Eloquent*, ed. Maxwell M. Wintrobe, 658–88. New York: McGraw-Hill.

Disput ob omolozhenii. 1924. *VM*, February 29, 2.

Dobronravov, S. 2001. "Bor'ba za zhiznesposobnost'": O filisofskikh istochnikakh meditsinskikh idei A. A. Bogdanova. *VMIAB* 8. http://www.bogdinst.ru/vestnik/vo8.htm (accessed September 12, 2004).

Donskov, S. I., and V. N. Iagodinskii. 2008. *Nasledie i posledovateli A. A. Bogdanova v sluzhbe krovi*. Moscow: n.p.

D-r S. Voronov ob omolozhenii liudei. 1924. *VM*, April 2, 2.

D-r Voronov sobiraetsia v Moskvu. 1925. *VM*, November 11, 2.

Elanskii, N. N. 1926. *Perelivanie krovi*. Moscow and Leningrad: GIZ.

———. 1927. A. Bogdanov, Bor'ba za zhiznesposobnost'. *ZhDUV* 7–8:595–96.

———. 1929. Perelivanie krovi v voennoi obstanovke. *NKhA* 17 (3): 426–47.

———. 1931. Voprosy metodiki i tekhniki perelivaniia krovi na fronte. *Voenno-sanitarnoe delo* 9:15–25.

Emel'ianov, A. Kh. 1928. Nasha organizatsiia donorstva. In Bogdanov, Bogomolets, and Konchalovskii 1928, 92–104.

English, Peter C. 1980. *Shock, physiological surgery, and George Washington Crile: Medical innovation in the progressive era.* Westport, CT: Greenwood Press.

Esakov, V. D. 1971. *Sovetskaia nauka v gody pervoi piatiletki.* Moscow: Nauka.

Fabrikant, M. B., and A. V. Taft. 1923. K voprosu ob omolozhenii (predvaritel'noe soobshchenie). *VD* 6–8:162–69.

Farr, A. D. 1979. Blood group serology: The first four decades, 1900–1939. *Medical History* 23:215–26.

Fedorovich, M. M., and A. F. Kaizer. 1925. K voprosu ob omolozhenii. *Turkestanskii meditsinskii zhurnal* 4 (9): 544–49.

Filatov, A. N. 1973. E. R. Gesse—odin iz osnovopolozhnikov transfuziologii v SSSR. *Vestnik khirurgii* 110 (6): 60–63.

Filomafitskii, A. M. 1848. *Traktat o perelivanii krovi (kak edinstvennom sredstve vo mnogikh sluchaiakh spasti ugasaiushchuiu zhizn'), sostavlennyi v istoricheskom, fiziologicheskom i khirurgicheskom otnoshenii.* Moscow: Universitetskaia tipografiia.

Fitzpatrick, Sheila. 1970. *The Commissariat of Enlightenment: Soviet organization of education and the arts under Lunacharsky.* Cambridge, England: Cambridge University Press.

———, ed. 1978. *Cultural revolution in Russia, 1928–1931.* Bloomington: Indiana University Press.

Fitzpatrick, Sheila, Alexander Rabinowitch, and Richard Stites, eds. 1991. *Russia in the era of NEP.* Bloomington: Indiana University Press.

Fleck, Ludwig. 1979. *Genesis and development of a scientific fact.* Trans. F. Bradley and T. J. Trenn. Ed. T. J. Trenn and R. K. Merton. Chicago: University of Chicago Press.

Freeman, Joseph T. 1979. *Aging's history and literature.* New York: Human Science Press.

Frenel', Diupiui de. 1924. *Perelivanie krovi.* Leningrad: GIZ.

Frenelle, Dupuy de. 1923. *La transfusion sanguine.* Paris: Edition Livres de France.

Fridman, M. P. 1923. K kazuistike peresadok zhelez vnutrennei sekretsii. *VD* 6–8:169–70.

Gaissinovitch, Abba E. 1985. Contradictory appraisal by K. A. Timiriazev of Mendelian principles and its subsequent perception. *History and Philosophy of the Life Sciences* 7:257–86.

Galton, Francis. 1870–71. Experiments in pangenesis, by breeding from rabbits of a pure variety, into whose circulation blood taken from other varieties had previously been largely transfused. *Proceedings of the Royal Society of London* 19:393–410.

Gare, Arran. 2000. Alexander Bogdanov's history, sociology and philosophy of science. *Studies in History and Philosophy of Science* 31 (2): 231–48.

Gaskins, Elizabeth. 1970. *The rise of experimental biology.* New York: Random House.

Gavrilov, O. K. 1968. *Ocherki istorii razvitiia i primeneniia perelivaniia krovi.* Leningrad: Meditsina.

Gess-de-Kal've, K. P. 1925. *Perelivanie krovi pri zlokachestvennykh novoobrazovaniiakh.* Kharkov: Nauchnaia mysl'.

153

Gess-de-Kal've, K. P., and G. Tutaev. 1924. O perelivanii krovi i vlivanii solevogo rastvora pri otravlenii asfikticheskimi iadami. *VD* 24–26:1335–41.

Gesse, E. R. 1925. O perelivanii krovi. *ZhDUV* 5:255–63.

———. 1926a. Organizatsiia professional'nogo donorstva v sviazi s operatsiei perelivaniia krovi. *Leningradskii meditsinskii zhurnal* 6:58–62.

———. 1926b. Pokazaniia k perelivaniiu krovi. *Vrachebnaia gazeta* 4:179–83; 5:227–30.

Getty, Arch. 1997. Pragmatists and puritans: The rise and fall of the Party Control Commission. The Carl Beck Papers in Russian and East European Studies, no. 1208.

Giangrande, Paul L. F. 2000. The history of blood transfusion. *British Journal of Haematology* 110:758–67.

Girgolav, S. S. 1922. Operatsiia "omolzheniia" po Steinach'u pri samoproizvol'nykh gangrenakh. *Sbornik nauchnykh trudov v chest' piatidesiatiletiia nauchno-vrachebnoi deiatel'nosti prof. A. A. Nechaeva*, 265–69. Petrograd: n.p.

Girón, Álvaro. 2003. Kropotkin between Lamarck and Darwin: The impossible synthesis. *Asclepio* 55:189–213.

Gliboff, Sander. 2006. The case of Paul Kammerer: Evolution and experimentation in the early 20th century. *JHB* 39:525–63.

Gloveli, Georgii. 1998. Bogdanov as scientist and utopian. In Biggart, Gloveli, and Yassour 1998, 40–59.

———. 2003. Strast' k monizmu: Gedonicheskii podbor Aleksandra Bogdanova. *VF* 9:110–27.

Golinsky, Jan. 1995. *Science as public culture: Chemistry and enlightment in Britian, 1760–1820*. Oxford, England: Oxford University Press.

Gorzka, Gabriele. 1980. *A. Bogdanov und der russische Proletkult: Theorie und Praxis einer sozialistischen Kulturrevolution*. Frankfurt: Campus.

Gottlieb, A. 1998. Karl Landsteiner, the melancholy genius: His time and his colleagues, 1868–1943. *Transfusion Medicine Reviews* 12 (1): 18–27.

Graham, Loren R. 1967. *The Soviet Academy of Sciences and the Communist Party, 1927–1932*. Princeton, NJ: Princeton University Press.

———. 1975. The formation of Soviet research institutes: A combination of revolutionary innovation and international borrowing. *Social Studies of Science* 5:303–29.

———. 1984a. Bogdanov's inner message. In Bogdanov 1984, 241–54.

———. 1984b. Bogdanov's *Red star*: An early Bolshevik science utopia. In *Nineteen eighty-four: Science between utopia and dystopia*, ed. Everett Mendelsohn and Helga Nowotny, 111–24. Dordrecht, Netherlands: Reidel.

———. 1987. *Science, philosophy and human behavior in the Soviet Union*. New York: Columbia University Press.

———, ed. 1990. *Science and the Soviet social order*. Cambridge, MA: Harvard University Press.

———. 1993. *Science in Russia and the Soviet Union: A short history*. Cambridge, England: Cambridge University Press.

———. 1999. *What have we learned about science and technology from the Russian experience?* Cambridge, MA: MIT Press.

Greene, Chas. W. 1922. The 1921 annual meeting of the Federation of American Societies for Experimental Biology. *Science* 55:379–80.

Greenfield, Douglas. 2006. Revenants and revolutionaries: Body and society in Bogdanov's Martian novels. *Slavic and East European Journal* 50 (4): 621–34.

Gremiatskii, M. A. 1925. *Chto takoe omolozhenie*. Moscow and Leningrad: GIZ.

Griffiths, John. 1980. *Three tomorrows: American, British and Soviet science fiction*. London: Macmillan.

Grille, Dietrich. 1966. *Lenins Rivale: Bogdanov und seine Philosophie*. Köln: Verlag Wissenschaft und Politik.

Gri-v, Gr. 1923. Bor'ba za dolgoletie i molodost'. *Gudok*, March 15, 4.

Gruman, Gerald J. 1966. A history of ideas about the prolongation of life: The evolution of prolongevity hypotheses to 1800. *Transactions of the American Philosophical Society* 56 (9): 1–102.

———. 2003. *A history of ideas about the prolongation of life*. New York: Springer.

Gudim-Levkovich, D. A. 1926. Nekotorye dannye o perelivanii krovi (tsitrathnoi). *Klinicheskaia meditsina* 2:45–47.

———. 1929a. Perelivanie krovi v obstanovke voennogo vremeni. *Voenno-sanitarnoe delo* 2:92–95.

———. 1929b. Tekhnika perelivaniia krovi. *Voenno-sanitarnoe delo* 3:42–51.

Gunson, Harold H., and Helen Dodsworth. 1996. Fifty years of blood transfusion. *Transfusion Medicine* 6:1–88. Suppl. no. 1.

Gusev, S. S. 1995. Ot "zhivogo opyta" k "organizatsionnoi nauke." In *Russkii pozitivizm: Lesevich, Iushkevich, Bogdanov*, ed. S. S. Gusev, 287–352. St. Petersburg: Nauka.

Hall, S. S. 2003. *Merchants of immortality: Chasing the dream of human life extension*. Boston: Houghton Mifflin.

Hamilton, David. 1986. *The monkey gland affair*. London: Chatto & Windus.

Hermann, Robert E. 1994. George Washington Crile (1864–1943). *Journal of Medical Biography* 2:78–83.

Hessen, Boris. 1931. The social and economic roots of Newton's *Principia*. In *Science at the crossroads*, 151–212. London: Cass & Co.

Holt, Niles R. 1971. Ernst Haeckel's monistic religion. *Journal of the History of Ideas* 32 (2): 265–80.

Howell, Yvonne. 2006. Eugenics, rejuvenation, and Bulgakov's journey into the heart of dogness. *Slavic Review* 65 (3): 544–62.

Huestis, Douglas W. 1996. The life and death of Alexander Bogdanov, physician. *Journal of Medical Biography* 4 (3): 141–47.

———. 2002. Alexander Bogdanov and medical science. In Bogdanov 2002b, 1–24.

———. 2007. Alexander Bogdanov: The forgotten pioneer of blood transfusion. *Transfusion Medicine Reviews* 21 (4): 337–40.

Hugh, Tomas. 2003. Early thoughts on death, disease, and sex. *International Journal of Epidemiology* 32 (1): 32–36.

Iagodinskii, V. N. 2006a. *A. A. Bogdanov: Bor'ba za zhiznesposobnost'*. Moscow: Terika.

———. 2006b. *Aleksandr Aleksandrovich Bogdanov*. Moscow: Nauka.

Iakov Moiseevich Bruskin (k 80-letiiu so dniia rozhdeniia). 1969. *Voprosy onkologii* 15 (5): 125–26.

Ikonnikov, S. N. 1971. *Sozdanie i deiatel'nost' ob"edinennykh organov TsKK—RKI v 1923-1934 gg*. Moscow: Nauka.

Institut perelivaniia krovi v Moskve. 1926. *VM*, April 3, 1.

Instruktsiia po primeneniiu lechebnogo metoda perelivaniia krovi. 1928. *Voprosy zdravookhraneniia* 17:71.

Ionsher, B. 1990. Utopia, fantastika, nauchnaia fantastika—k poniatiinoi sisteme v russkoi sovetskoi literature 20-ikh gg. *Zeitschrift fur Slawistik* 35 (3): 360–65.

Isaev, V. 1922. Problema omolozheniia. *ChiP* 4:37–64.

Iul'ev, P. 1923. Omolazhivanie. *Gigiena i zdorov'e rabochei sem'i* 1:1–3.

Ivanova, A. T. 1972. *Sergei Petrovich Fedorov (1869–1936): Nauchnaia biografiia.* Moscow: Meditsina.

Jacobs, Walter D. 1970. *Frunze: The Soviet Clausewitz, 1885–1925.* The Hague: M. Nijhoff.

Jansky, J. 1906–7. Haematologicke studie u psychotiku. *Sbornik Klinicky* (Praha) 8:85–139.

Jaworski, Helán. 1925a. La question du rajeunissement. *Revue scientifique* 17:599–600.

———. 1925b. *La régénération de l'organisme humain par les injections de sang.* Paris: Maloine.

———. 1926. *Pourquoi la mort? L'intériorisation.* Paris: J. Oliven.

Jensen, K. M. 1978. *Beyond Marx and Mach: Aleksandr Bogdanov's philosophy of living experience.* Dordrecht: Reidel.

———. 1982. Red star: Bogdanov builds a utopia. *Studies in Soviet Thought* 23 (1): 1–34.

Joravsky, David. 1961. *Soviet Marxism and natural science, 1917–1932.* New York: Columbia University Press.

Kammerer, Paul. 1921. Über Verjüungun und Verlängerung des persönlichen Lebens: Die Versuche an Pflanze, Tier und Mensch gemeinverständlich dargestellt. Stutgart and Berlin: DVA.

———. 1922a. *Omolazhivanie i dolgovechnost'.* Petrograd and Berlin: Z. I. Grzhebin.

———. 1922b. *Omolozhenie i prodlenie lichnoi zhizni.* Moscow: GIZ.

Katz, Stephen. 1996. *Disciplining old age: The formation of gerontological knowledge.* Charlottesville: University Press of Virginia.

Kavtaradze, P. P. 1960. *Zhizn' i deiatel'nost' zasluzhennogo professora Ia. A. Anfimova.* Tbilisi: Sabchota Sakartvelo.

Kenig, E. I. 1936. *Mezhdunarodnaia bibliografiia po voprosam perelivaniia krovi i ucheniiu o krovianykh gruppakh za 1900–1933 gg.* Leningrad: Vestnik khirurgii.

Keynes, Geoffrey. 1922. *Blood transfusion.* London: Holder & Stoughton.

Keynes, John Maynard. 1920. *The economic consequences of the peace.* New York: Harcourt, Brace and Howe.

Khakhordin, Oleg. 1999. *The collective and the individual in Russia: A study of practices.* Berkeley: University of California Press.

Kharin, Iu. M. 1924. Bor'ba so starost'iu. *Molodaia gvardiia* 7:224–39.

Khromov, S. S. 2001. *Leonid Krasin: Neizvestnye stranitsy biografii, 1920–1926 gg.* Moscow: RAN-Institut Istorii.

Kino. Beseda s professorom Kol'tsovym o fil'me "Omolozhenie." 1923. *Izvestiia*, October 1, 5.

Kino, Kul'tob"edinenie. 1924. *Pravda*, September 7, 5.

K istorii bolezni tov. Frunze. 1925. *Pravda*, November 3, 3.

Knopp, Shoshana. 1985. Herbert Spencer in Čexov's "Skučnaja istorija" and "Duél": The love of science and the science of love. *Slavic and East European Journal* 29 (3): 279–96.

Koeniker, D. P., W. G. Rosenberg, and R. G. Suny, eds. 1989. *Party, state, and society in the Russian civil war.* Bloomington: Indiana University Press.

Koestler, Arthur. 1971. *The case of the midwife-toad.* New York: Random House.

Kogan, I. G. 1924. Peresadka polovykh zhelez u zhivotnykh i cheloveka. *Biulleten' MOIP* (otdel eksperimental'noi biologii) 1:179–95.

Kogo nado omolazhivat'. 1925. *VM*, September 3, 3.

Kogo nuzhno omolazhivat'. 1925. *VM*, October 29, 3.

Kohler, Robert E. 1991. *Partners in science: Foundations and natural scientists, 1900–1945.* Chicago: University of Chicago Press.

———. 1994. *Lords of the fly: Drosophila genetics and laboratory life.* Chicago: University of Chicago Press.

———. 2003. *From medical chemistry to biochemistry: The making of a biomedical discipline.* New York: Cambridge University Press.

Kol'tsov, N. K. 1923a. Novaia literatura po omolozheniiu. *Uspekhi eksperimental'noi biologii* 2 (1–2): 140–44.

———, ed. 1923b. *Omolozhenie.* Moscow and Petrograd: n.p.

———. 1923c. Omolozhenie organizma. *Pravda*, November 14, 2–3.

———. 1924a. Noveishaia amerikanskaia literatura v oblasti operativnovo omolozheniia cheloveka. In Kol'tsov 1924b, 124–47.

———, ed. 1924b. *Omolozhenie.* Moscow and Petrograd: GIZ.

———. 1925. Chudesnye dostizheniia nauki. *Narodnyi uchitel'* 4:61–75.

Komissiia po organizatsii pokhoron A. A. Bogdanova. 1928. *Pravda*, April 8, 5.

Konchalovskii, M. P. 1928. Bolezn' i smert' A. A. Bogdanova. In Bogdanov, Bogomolets, and Konchalovskii 1928, XII–XXV.

———. 1931. Perelivanie krovi kak lechebnyi metod. In A. Pinei, *Poslednie dostizheniia gematologii*, 242–51. Moscow and Leningrad: GIZ.

Korganova-Miuller, F. S. 1925. K voprosu o prichinakh reaktsii posle perelivaniia krovi. *Russkaia klinika* 4 (15): 46–59.

Korschelt, Eugene. 1922. *Lebensdauer, Altern und Tod.* Jena: Fischer.

Korzhikhina, T. P. 1994. *Sovetskoe gosudarstvo i ego uchrezhdeniia.* Moscow: RGGU.

Kostiukov, M. Kh. 1927. K izucheniiu gomoplastiki s trupa: Kak dolgo ostaiutsia tkani steril'nymi posle smerti. *NKh* 4 (1): 3–9.

Koziner, V. B. 1987. A. A. Bogomolets i evoliutsiia vzglaidov na deistvie perelitoi krovi. *Gematologiia i transfuziologiia* 5:3–8.

Kozlov, B. I., and G. A. Savina. 2008. *Kommunisticheskaia akademiia TsIK SSSR (1918–1936).* Moscow: Slovo.

Kramarenko, E. Iu., and L. A. Barinshtein. 1924. Nabor dlia perelivaniia krovi. *Sovremennaia meditsina* 7–9:56–60.

Krasil'nikov, S. A., K. N. Morozov, and I. V. Chubykin, eds. 2002. *Sudebnyi protsess nad sotsialistami-revoliutsionerami (iiun'-avgust 1922): Podgotovka. Provedenie. Itogi. Sbornik dokumentov.* Moscow: ROSSPEN.

Krasin, L. B. 2002. *Pis'ma k zhene i detiam, 1917–1926.* Ed. Iu. G. Fel'shtinskii, G. I. Cherniavskii, and F. Markiz. http://kulichki.com/moshkow/history/felshtinsky/Krasin.Pisma.txt (accessed September 21, 2005).

Kratkii otchet o zasedaniiakh khirurgicheskoi sektsii 3-go nauchnogo kongressa vrachei Gruzii v Tiflise 24-28 maia 1925. 1925. *NKhA* 8 (3): 444-46.

Krementsov, Nikolai. 1997. *Stalinist science*. Princeton, NJ: Princeton University Press.

———. 2005. *International science between the world wars: The case of genetics*. London: Routledge.

———. 2006. Big revolution, little revolution: Science and politics in Bolshevik Russia. *Social Research* 73 (4): 1173-1204.

———. 2008. Hormones and the Bolsheviks: From organotherapy to experimental endocrinology, 1918-1929. *Isis* 99 (3): 486-518.

———. 2009. Off with your heads: Isolated organs in early Soviet science and fiction. *Studies in History and Philosophy of Biological and Biomedical Sciences* 40 (2): 87-100.

———. 2010. Marxism, Darwinism, and genetics in Soviet Russia. In *Biology and Ideology: From Descartes to Dawkins*, ed. Ron Numbers and Denis Alexander, 215-46. Chicago: University of Chicago Press.

Kremlevskaia meditsina (ot istokov do nashikh dnei). 1997. Moscow: Izvestiia.

Krinitskii, Ia. M., and Kh. Kh. Rutkovskii. 1927. *Perelivanie krovi: Dlia vrachei i studentov*. Leningrad: Obrazovanie.

Kropotkin, Petr. 1902. *Mutual aid: A factor of evolution*. London: Heinemann.

———. 1904. *Vzaimnaia pomoshch' sredi liudei i zhivotnykh*. Moscow: M. D. Orekhov.

Kuda ustraivat' ekskursii. 1924. *Krasnyi medrabotnik*, December 15, 4.

Kustria, D. K. 1924. Sluchai poiavleniia zubov u starogo kota, omolozhennogo posredstvom transplantatsii semennikov. *Russkii fiziologicheskii zhurnal* 7 (1-6): 361-62.

Landsteiner, K. 1901. Uber Agglutinationserscheinungen normalen mensclinchen Blutes. *Wiener Klinik Wochenschtire* 1:5-8.

Latour, Bruno, and Steve Woolgar. 1976. *Laboratory life: The social construction of scientific facts*. Princeton, NJ: Princeton University Press.

Lecourt, Dominique. 1976. *Lyssenko: Histoire réelle d une science prolétarienne*. Paris: François Maspéro.

———. 1977. *Proletarian science? A case of Lysenko*. London: NLM.

Lederer, Susan. 2008. *Flesh and blood: Organ transplantation and blood transfusion in twentieth-century America*. New York: Oxford University Press.

Lenin, V. I. [Vl. Il'in]. 1909. *Materializm i empiriokrititsizm: Kriticheskie zametki ob odnoi reaktsionnoi filosofii*. Moscow: Zveno.

——— [N. Lenin]. 1920. *Materializm i empiriokrititsizm: Kriticheskie zametki ob odnoi reaktsionnoi filosofii*. Moscow: GIZ.

Lenin, V. I., and G. V. Plekhanov. 1923. *Protiv Bogdanova*. Moscow: Krasnaia nov'.

Lincoln, W. B. 1989. *Red victory: A history of Russian civil war, 1918-1921*. New York: Da Capo Press.

Liubutin, K. N. 2003. Rossiiskie versii filosofii marksizma: Aleksandr Bogdanov. *VF* 9:76-91.

London, E. S., and I. I. Kryzhanovskii. 1924. *Bor'ba za dologovechnost'*. Petrograd: Put' k znaniiu.

Lubrano, Linda L., and Susan G. Solomon, eds. 1980. *The social context of Soviet science*. Boulder, CO: Westview Press.

Lunacharski, A. 1928. Aleksandr Aleksandrovich Bogdanov. *Pravda*, April 10, 5.

Lustig, A. J. 2000. Sex, death, and evolution in proto- and metazoa, 1876–1913. *JHB* 33:221–46.

Lutsenko, A. V. 2003. Nachalo konflikta mezhdu V. I. Leninym i A. A. Bogdanovym (1907–1909 gg.). *Voprosy istorii* 1:28–47.

Maiants, I. A. 1925. O perelivanii krovi v akushersko-ginekologicheskoi i khirurgicheskoi praktike. *Zhurnal akusherstva i ginekologii* 36 (5): 693–95.

Mally, Lynn. 1990. *Culture of the future: The Proletkult movement in revolutionary Russia*. Berkeley: University of California Press.

Maloletkov, S. 1929. Krovianye gruppy i ikh opredelenie. *Voenno-sanitarnoe delo* 2:96–103.

Mandel'shtam, A. E. 1925. K voprosu o perelivanii krovi (Novyi prostoi sposob opredeleniia izogemoaggliutinatsii). *ZhDUV* 5:263–66.

Mänicke-Gyöngyösi, Krisztina. 1982. "Proletarische Wissenschaft" und "sozialistische Menschheitsreligion" als Modelle proletarischer Kultur: Zur linksbolschewistischen Revolutionstheorie A. A. Bogdanovs und A. V. Lunačarskijs. Berlin: Osteuropa-Institut an der Freien Universität.

McGuire, Patrick L. 1985. *Red stars: Political aspects of Soviet science fiction*. Ann Arbor, MI: UMI Research Press.

Mendelsohn, Everett, Peter Weingart, and Richard Whitley, eds. 1977. *The social production of scientific knowledge*. Dordrecht: D. Reidel.

Metchnikoff, Elie. 1901-2. Etudes biologiques sur la vieillesse. *Annales de l'Institut Pasteur* 15:865–79; 16:912–17.

———. 1903. *Etudes sur la nature humaine: Essai de philosophie optimiste*. Paris: Masson & C-ie.

Miasnikov, L. N. 1999. Obshchii iazyk v utopii. *Chelovek* 4–5. http://vivovoco.rsl.ru/vv/papers/men/utolang.htm (accessed September 5, 2006).

Mikhel', D. V. 2006. Perelivanie krovi v Rossii, 1900–1940. *Voprosy istorii estestvoznaniia i tekhniki* 2:99–113.

Moore, Pete. 2003. *Blood and justice*. Chichester, West Sussex, England: Wiley.

Moss, William L. 1910. Studies of isoagglutinins and isohemolysins. *Bulletin of the Johns Hopkins Hospital* 21:63–70.

Müller-Wille, Staffan, and Hans-Jörg Rheinberger, eds. 2007. *Heredity produced: At the crossroads of biology, politics, and culture, 1500–1870*. Cambridge, MA: MIT Press.

Naiman, Eric. 2002. Injecting Communism: A. A. Zamkov, Soviet endocrinology and the Stalinist body. http://www.virginia.edu/crees/Naiman%20paper.pdf (accessed March 6, 2003).

Nathoo, N., F. K. Lautzenheiser, and G. H. Barnett. 2009. The first direct human blood transfusion: The forgotten legacy of George W. Crile. *Neurosurgery* 64 (3): ons20–ons27.

Nauka vyshe boga. 1925. *VM*, June 20, 2.

Nechai, A. I. 1977. *V. N. Shamov*. Moscow: Meditsina.

Neiman, F. T. 1925. O tekhnike predvaritel'nykh reaktsii pri perelivanii krovi. *Ekaterinoslavskii meditsinskii zhurnal* 3–4:200–202.

Nemilov, A. V. 1923a. Noveishie opyty omolazhivaniia liudei. *ChiP* 4–5:25–30.

———. 1923b. O peresadke semennykh zhelez u mlekopitaiushchikh i cheloveka. *Priroda* 7–12:78–84.

———. 1925. "Omolozhenie" sel'sko-khoziastvennykh zhivotnykh. *ChiP* 1:23–36.

Ne omolozhenie, a osvezhenie. 1924. *VM*, March 4, 2.

Nevskii, V. 1920. Dialekticheskii materializm i filosofia mertvoi reaktsii. In Lenin 1920, 369–84.

N.N. 1927. A. Bogdanov, Bor'ba za zhiznesposobnost'. *VM*, January 28, 3.

Novachenko, N. P., ed. 1968. *Sbornik nauchnykh trudov po khirurgii i neirokhirurgii, posviashchennyi 50-letnei deiatel'nosti professora V. N. Shamova*. Leningrad and Khar'kov: n.p.

Novoselov, V. I. 1994. *Marsiane iz pod Vologdy*. Vologda: Ardvisura.

———. 2004. Vologodskaia ssylka A. A. Bogdanova. *VMIAB* 17. http://www.bogdinst.ru/vestnik/v17.htm (accessed September 23, 2006).

Novosti nauki i tekhniki: Dostizheniia russkoi nauki. 1924. *Izvestiia*, December 16, 6.

Novyi sposob perelivaniia krovi (beseda s privat-dotsentom 1 MGU doktorom Ia. M. Bruskinym). 1926. *VM*, January 6, 2.

Nudel'man, R. 1966. Fantastika, rozhdennaia revoliutsiei. In *Fantastika*, vol. 3, 330–69. Moscow: Molodaia gvardiia.

Oblastnoi s"ezd khirurgov levoberezhnoi Ukrainy. 1925. *NKhA* 8 (1): 152–56.

Ocherki filosofii kollektivizma. 1909. St. Petersburg: Znanie.

O'Connor, Timothy E. 1983. *The politics of Soviet culture: Anatolii Lunacharskii*. Ann Arbor, MI: UMI Research Press.

———. 1992. *The engineer of revolution: L. B. Krasin and the Bolsheviks*. Boulder, CO: Westview Press.

Odom, William E. 1973. *The Soviet volunteers: Modernization and bureaucracy in a mass public organization*. Princeton, NJ: Princeton University Press.

"O kadrakh donorov": Postanovlenie SNK RSFSR ot 22 aprelia 1935. 1935. *Sobranie uzakonenii i rasporiazhenii raboche-krest'ianskogo pravitel'stva RSFSR* 12:132.

Olby, Robert C. 1966. *Origins of Mendelism*. New York: Schoken Books.

Oleinik, S. F. 1955. *Perelivanie krovi v Rossii i SSSR*. Kiev: Meditsinskoe izdatel'stvo UkSSR.

Omolozhenie. 1923a. *Izvestiia*, October 9, 4.

Omolozhenie. 1923b. *V masterskoi prirody* 7:64.

Omolozhenie liudei (beseda s prof. Zavadovskim). 1924. *VM*, March 26, 2.

Omolozhenie liudei i zhivotnykh. 1923. *Izvestiia*, November 20, 5.

Omolozhenie Lloid-Dzhorzha. 1924. *VM*, September 10, 2.

Omolozhenie v Rossii. 1924. Leningrad: Meditsina.

Omolozhenie zhivotnykh: Novaia stat'ia dokhoda dlia predpriimchivykh kapitalistov. 1925. *VM*, March 31, 2.

O perelivanii krovi. 1923a. *Gigiena i zdorov'e rabochei sem'i* 5:15.

O perelivanii krovi. 1923b. *NiT* 39:14.

Oppel', V. 1930. Perelivanie krovi. *Gigiena i zdorov'e rabochei i krest'ianskoi sem'i* 4:11–12; 5:5; 6:13–14.

Opyt omolozheniia. 1923. *Izvestiia*, December 1, 4.

Opyty professora Voronova. 1922. *Izvestiia*, October 31, 5.

O tak-nazyvaemom omolozhenii. 1923. *Izvestiia*, March 4, 4.

Otkrytie instituta po perelivaniiu krovi. 1926. *VM*, April 1, 2.

Otkrytie Mezhdunarodnogo Instituta A. Bogdanova. 2000. *VMIAB* 1. http://www.bogdinst.ru/vestnik/v01_01.htm (accessed September 12, 2004).

Ottenberg, N. 1926. A. Bogdanov, Bor'ba za zhiznesposobnost'. *Izvestiia*, December 28, 5.

Oudsboorn, Nelly. 1994. *Beyond the natural body: An archeology of sex hormones*. London: Routledge.

Palladin, A. V. 1923. *Omolozhenie.* Khar'kov: Put' prosveshcheniia.

Pamiati A. A. Bogdanova. 1928a. *Izvestiia,* April 8, 2.

Pamiati A. A. Bogdanova. 1928b. *Izvestiia,* April 28, 7.

Pamiati A. A. Bogdanova. 1928c. *Izvestiia,* May 6, 4.

Pauchet, Victor, and Auguste Bécart. 1924. *La transfusion du sang.* Paris: G. Doin.

Pauly, Philip J. 1987. *Controlling life: Jacques Loeb and the engineering ideal in biology.* New York: Oxford University Press.

Pearl, Raymond. 1922. *The biology of death.* Philadelphia and London: Lippincott.

Pelis, Kim. 1997. Blood clots: The nineteenth century debate over the substance and means of transfusion in Britain. *Annals of Science* 54:331–60.

———. 1999. Transfusion with teeth. In *Manifesting medicine: Bodies and machines,* ed. Robert Bud, Bernard Finn, and Helmuth Trischler, 1–29. Amsterdam: Harwood.

———. 2001a. Blood standards and failed fluids: Clinic, lab, and transfusion solutions in London, 1868–1916. *History of Science* 39/2 (124): 185–213.

———. 2001b. Taking credit: The Canadian Army Medical Corps and the British conversion to blood transfusion in WWI. *Journal of the History of Medicine and Allied Sciences* 56 (3): 238–77.

———. 2002. Edward Archibald's notes on blood transfusion in war surgery—A commentary. *Wilderness and Environmental Medicine* 13 (3): 211–14.

Pereimenovanie Instituta perelivaniia krovi v Institut perelivaniia krovi im. Bogdanova. 1928. *Pravda,* April 18, 7.

Perelivanie krovi. 1923. *NiT* 39:14.

Perelivanie krovi. 1924. *Izvestiia,* April 16, 6.

Perelivanie krovi. 1925a. *Izvestiia,* October 23, 4.

Perelivanie krovi. 1925b. *NiT* 22:21.

Perelivanie krovi. 1925c. *VM,* October 20, 3.

Pertsev, V. A. 2003. Perelivanie krovi. *Novyi khirurgicheskii arkhiv* (Internet Journal) 2 (3). http://www.surginet.info/nsa/2/3/index.html (accessed November 16, 2006).

Pervye itogi rabot prof. Voronova. 1925. *VM,* December 3, 3.

Pickering, Andrew R., ed. 1992. *Science as practice and culture.* Chicago: University of Chicago Press.

Pickstone, John. 2001. *Ways of knowing: A new history of science, technology and medicine.* Chicago: University of Chicago Press.

Pis'mo TsK RKP(b) "O Proletkul'takh." 1920. *Pravda,* December 1, 1.

Platonov, G. V. 2001. *Kliment Arkadyevich Timiryazev.* Honolulu: University Press of the Pacific.

Plekhanov, G. V. 1925. *Protiv Bogdanovshchiny.* Kharkov: Proletarii.

Pokhod protiv omolozheniia. 1925. *VM,* September 5, 3.

Pokhorony A. A. Bogdanova. 1928. *Pravda,* April 11, 5.

Pokhorony A. A. Bogdanova (Malinovskogo). 1928. *Izvestiia,* April 11, 4.

Pokotilo, V. 1927. Ia. M. Bruskin, Perelivanie krovi. *NKh* 4 (3): 315–18.

Politbiuro TsK RKP(b)-VKP(b): Povestki dnia zasedanii, 1919–1952. 2000. 3 vols. Moscow: ROSSPEN.

Poltavskii, Il. 1928. Krov'. *VM,* February 2, 2.

Ponomarev, B. N., ed. 1973. *Istoriia Kommunisticheskoi Partii Sovetskogo Soiuza.* Moscow: Politizdat.

Posle omolozheniia: O konchine Rotshilda. 1925. *VM,* April 14, 3.

161

Poustilnik, Simona. 1998. Biological ideas in tektology. In Biggart, Gloveli, and Yassour 1998, 63–73.

Predstavliaet li opasnost' perelivanie krovi? 1928. *VM*, April 10, 1.

Prodazhnaia tsena krovi. 1925. *VM*, June 4, 3.

Prot'ko, T. S., and A. A. Gratsianov. 2010. *Aleksandr Bogdanov*. Moscow: Knizhnyi dom.

Protokol vskrytiia. 1925. *Pravda*, November 1, 1.

Protokol zasedaniia. 1926. *Vestnik khirurgii i pogranichnykh oblastei* 6 (17–18): 223–24.

Pyzhov, N. 1923. Problema omolozheniia. *Krasnaia niva* 48:26.

Rafal'kes, S. 1926. Sovremnnoe sostoianie voprosa o gruppakh chelovecheskoi krovi. *VSM* 9:8–13.

———. 1927. Sushchestvuet-li russkaia nauka? *Kazanskii meditsinkii zhurnal* 3:361–63.

Rakovskii, Kh. 1926. Leonid Borisovich Krasin. *Vestnik Komakademii* 18:5–16.

Rees, E. A. 1987. *State control in Soviet Russia: The rise and fall of the Workers' and Peasants' Inspectorate, 1920–1934*. Basingstoke, England: Macmillan.

Reiter, Wolfgang L. 1999. Zerstort und vergessen: Die Biologische Versuchsanstalt und ihre Wissenschaftler/innen. *Osterreichischezeitschrift fur Geschichtswissenschaften* 10 (4): 585–614.

Revich, Vsevolod. 1985. Perekrestok utopii: U istokov sovetskoi fantastiki. In *Orion*, ed. N. Berkova, 309–48. Moscow: Moskovskii rabochii.

Richards, Robert J. 2008. *The tragic sense of life: Ernst Haeckel and the struggle over evolutionary thought*. Chicago: University of Chicago Press.

Richter, Jochen. 2007. Pantheon of brains: The Moscow Brain Research Institute, 1925–1936. *Journal of the History of the Neurosciences* 16:138–49.

Robertson, T. Brailsford. 1923. *The chemical basis of growth and senescence*. Philadelphia and London: J. B. Lippincott.

Rodionov, F. I. 1923. Operatsiia Shteinakha pri starcheskom marazme. *Vrachebnaia gazeta* 23:499.

———. 1924. Shteinakhovskaia operatsiia u bol'noi s arteriosklerot01cheskim psikhozom. *Zhurnal psikhologii, nevrologii i psikhiatrii* 4:144–51.

Rogachevskii, Andrei, and Milena Michalski. 1994. Social Democratic Party schools on Capri and in Bologna in the correspondence between A. A. Bogdanov and A. V. Amfiteatrov. *Slavonic and East European Review* 72 (4): 664–79.

Rubashkin, V. Ia. 1929. *Krovianye gruppy*. Moscow and Leningrad: GIZ.

Rufanov, I. G. 1927a. A. Bogdanov, Bor'ba za zhiznesposobnost'. *Russkaia klinika* 7 (34): 314.

———. 1927b. Bruskin, Ia. M. Perelivanie krovi. *Russkaia klinika* 7 (34): 312–13.

———. 1927c. Elanskii, N. N. Perelivanie krovi. *Russkaia klinika* 7 (34): 309–10.

Rullkötter, Bernd. 1974. *Die wissenschaftliche Phantastik der Sowjetunion: Eine vergleichende Untersuchung der spekulativen Literatur in Ost und West*. Bern: H. Lang.

Ryzhkov, V. 1923. Bor'ba so starost'iu. *Znanie* (Khar'kov) 4:27–30.

Sadovskii, V. N. 2003a. Empiriomonizm A. A. Bogdanova: Opyt prochteniia spustia stoletie posle publikatsii. *VF* 9:92–109.

———. 2003b. Istoriia sozdaniia, teoreticheskie osnovy i sud'ba empiriomonizma A. A. Bogdanova. In Bogdanov 2003a, 340–65.

Samoomolozhenie d-ra Voronova. 1924. *VM*, May 9, 2.

Schlich, Thomas. 1996. "Welche Macht uber Tod und Leben!" Die Etablierung der Bluttransfusion im Ersten Weltkrieg. In *Die Medizin und Der Erster Weltkrieg*, ed. Wolfgang U. Eckart and Christoph Gradmann, 109–30. Pfaffenweiler: Centaurus-Verlagsgesellschaft.

Schmidt, Paul J. 2009. Transfusion in 1935 at the first ISBT Congress. *Transfusion Today* 81:9–10.

Schmidt, Peter. 1922. *Theorie und Praxis der Steinach'schen Operation*. Wien, Leipzig, and München: Rikola Verlag.

———— [Shmidt]. 1923. *Teoriia i praktika omolozheniia (operatsii Shteinakha)*. Petrograd: Prakticheskaia meditsina.

Schneider, William H. 1983. Chance and social setting in the application of the discovery of blood groups. *Bulletin of the History of Medicine* 57:545–62.

————. 1995. Blood group research in Great Britain, France and the United States between the world wars. *Yearbook of Physical Anthropology* 38:77–104.

————, ed. 1996. *The first genetic marker: Blood group research, race and disease, 1900–1950*. Special issue, *History and Philosophy of the Life Sciences* 18 (3): 273–369.

————. 1997. Blood transfusion in peace and war. *Social History of Medicine* 10 (1): 105–26.

————. 2003. Blood transfusion between the wars. *Journal of the History of Medicine and Allied Sciences* 58 (2): 187–224.

Semashko, N. A. 1923. Okhrana zdorov'ia kommunistov. *Izvestiia*, February 16, 3.

————. 1928a. Smert' A. A. Bogdanova. *Izvestiia*, April 8, 2.

————. 1928b. Smert' A. A. Bogdanova (Malinovskogo). *Pravda*, April 8, 5.

Sengoopta, Chandak. 2006. *The most secret quintessence of life: Sex, glands, and hormones, 1850–1950*. Chicago: University of Chicago Press.

Shamov, V. N. 1911. *O znachenii fizicheskikh metodov dlia khirurgii zlokachestvennykh novoobrazovanii*. St. Petersburg: Iu. N. Erlikh.

————. 1914a. Iz khirurgicheskikh nabliudenii v Angliii i Severnoi Amerike. *Voenno-meditsinskii zhurnal* 240 (7): 400–423; (8): 9–26.

————. 1914b. Podgotovka vrachei v amerikanskoi armii. *Voenno-meditsinskii zhurnal* 240 (11): 518–40.

————. 1915. The Rockefeller Institute for Medical Research. *Trudy gospital'noi khirurgicheskoi kliniki prof. S. P. Fedorova* 9:1–94.

———— [Shamoff]. 1915–16a. Concerning the action of various pituitary extracts upon the isolated intestinal loop. *American Journal of Physiology* 39:268–78.

———— [Shamoff]. 1915–16b. On the secretory discharge of the pituitary body produced by stimulation of the superior cervical sympathetic ganglion. *American Journal of Physiology* 39:279–90.

————. 1921. O perelivanii krovi. *NKhA* 1 (1): 21–27.

————. 1931. Perelivanie krovi v SSSR i klinicheskoe znachenie etogo metoda. In *Trudy 4-go Vseukrainskogo s"ezda khirurgov*. Published as a special issue of *NKhA* 23 (3–4): 9–18.

————. 1937. Moi dvadtsatiletnii opyt po perelivaniiu krovi. *VD* 11:787–802.

Shamov, V. N., and N. N. Elanskii. 1923. Izoaggliutiniruiushchie svoistva chelovecheskoi krovi, znachenie ikh dlia khirurgii i sposoby opredeleniia. *NKhA* 3 (11): 565–95.

Shamov, V. N., and M. Kh. Kostiukov. 1929. K izucheniiu gomoplastiki s trupa—perelivanie krovi ot trupa. *Trudy 3-go Vseukrainskogo s"ezda khirurgov v*

Dnepropetrovske, 9–14 sentiabria 1928. Published as a special issue of NKhA 18 (1–3): 184–90.

Sharapov, Iu. P. 1997. Lenin i Bogdanov: Ot sotrudnichestva k protivostoianiiu. *Otechestvennaia istoriia* 5:55–67.

Shcheglov, A. V. 1937. *Bor'ba Lenina protiv bogdanovskoi revizii marksizma.* Moscow: Sotsekgiz.

Shilovtsev, S. P. 1923. Noveishie dannye o perelivanii krovi. *Saratovskii vestnik zdravookhraneniia* 7:55–57.

Shirokogorov, I. I. 1924. Kritika teorii Shteinakha. *Zhurnal teoreticheskoi i prakticheskoi meditsiny* 1 (1–2): 1–17.

Shmidt, P. Iu. 1923. *Omolazhivanie.* Petrograd: Soikin.

———. 1924. *Bor'ba so starost'iu.* Leningrad: GIZ.

———. 1925. Smert' i bessmertie s biologicheskoi tochki zreniia. *Novyi mir* 5:130–41.

Shtark, V. A. 1923. *Omolozhenie: Popytka resheniia problemy vechnoi molodosti.* Baku: n.p.

Shushpanov, A. N. 2001. Osobennosti razvitiia zhanra otechestvennoi utopii v 1920-e gg. *VMIAB* 9. http://www.bogdinst.ru/vestnik/doc09/05.doc (accessed September 12, 2004).

Shustrov, N. M. 1924. Transplantatsiia polovykh zhelez. *VD* 20–23:1210–14.

Shustrov, N. M., and I. G. Dmitriev. 1923. O zhiznennosti eritrotsitov pri perelivanii krovi. *NKhA* 3 (9): 49–57.

Shustrov, N. M., S. G. Karpova, and I. V. Tikhomirov. 1924. Opyty nad perezhivaniem i peresadkoi polovykh zhelez. *VD* 20–23:1133–35.

Simonov, V. V., and N. K. Figurovskaia. 1988. Tri utopii Aleksandra Bogdanova. In *Sotsiokul'turnye utopii XX veka* (Moscow: INION) 6:71–96.

Sirotinin, N. N. 1967. *A. A. Bogomolets.* Moscow: Meditsina.

Skundina, M. G. 1935. Osnovnye etapy razvitiia problemy perelivaniia trupnoi krovi. *Sovetskaia khirurgiia* 5:69–77.

Smit, M., and A. Timiriazev, eds. 1925. *Statisticheskii metod v nauchnykh issledovaniiakh.* Moscow: Kommunisticheskai akademiia.

Sobolev, I. I. 1992. Tovarishch. In *Trudy komissii po nauchnomu naslediiu A. A. Bogdanova,* 218–22. Moscow: Institut ekonomiki RAN.

Soboleva, Maja. 2007. *Aleksandr Bogdanov und der philosophische Diskurs in Russland zu Beginn des 20. Jahrhunderts: Zur Geschichte des russischen Positivismus.* Hildesheim and New York: G. Olms.

Sochor, Zenovia. 1988. *Revolution and culture: The Bogdanov-Lenin controversy.* Ithaca, NY: Cornell University Press.

Sokolov, N. 1927. A. Bogdanov, Bor'ba za zhiznesposobnost'. *Kazanskii meditsinskii zhurnal* 3:361.

Solomon, Susan G., and John F. Hutchinson, eds. 1990. *Health and society in revolutionary Russia.* Bloomington: Indiana University Press.

Sostoianie zdorov'ia L. B. Krasina. 1925. *VM,* November 23, 1.

Sostoianie zdorov'ia narkomvoenmora i predsedatelia RVS SSSR tov. M. V. Frunze. 1925. *Pravda,* October 29, 3.

Spasokukotskii, S. I. 1931. Voprosy, sviazannue s perelivaniem krovi v obstanovke voennogo vremeni. *NKh* 12 (1): 64–70.

Starr, Douglas P. 1998. *Blood: An epic history of medicine and commerce.* New York: Knopf.

Steinach, Eugen. 1920. Verjüngung durch experimentelle Neubelebung der alternden Pubertätsdrüse. *Roux' Archiv für Entwicklungsmechanik* 46:557–618.

Stites, Richard. 1984. Fantasy and revolution: Alexander Bogdanov and the origins of Bolshevik science fiction. In Bogdanov 1984, 1–16.

———. 1989. *Revolutionary dreams: Utopian vision and experimental life in the Russian Revolution*. New York: Oxford University Press.

Stone, David R. 2000. *Hammer and rifle: The militarization of the Soviet Union, 1926–1933*. Lawrence: University Press of Kanzas.

Sukhov, V. N. 1925. Ob operatsii omolozheniia po Steinach'u. *Leningradskii meditsinskii zhurnal* 1:24–28.

Sutugin, V. V. 1865. *O perelivanii krovi*. St. Petersburg: Ia. Trei.

Suvin, Darko. 1971. The utopian tradition of Russian science fiction. *Modern Language Review* 66 (1): 139–59.

———. 1979. *Metamorphoses of science fiction: On the poetics and history of a literary genre*. New Haven, CT: Yale University Press.

S.V. 1922. Omolazhivanie liudei. *Poslednie novosti*, July 10, 1.

Tarasevich, L. A., and V. A. Liubarskii, eds. 1924. *Gosudarstvennyi institut narodnogo zdravookhraneniia imeni Pastera ("GINZ"), 1919–1924*. Moscow: GINZ.

Tartarin, Robert. 1994. Transfusion sanguine et immortalité chez Alexandr Bogdanov. *Droit et Société* 28:565–81.

Teoriia prof. Shteinakha v kino. 1922. *Poslednie novosti*, August 21, 3.

Timiriazev, K. A. 1892–95. Istoricheskii metod v biologii. *Russkaia mysl'* 8 (1892): 83–99; 10 (1892): 142–63; 8 (1893): 39–58; 6 (1894): 81–95; 7 (1894): 90–101; 7 (1895): 73–89.

———. 1894a. *Charlz Darvin i ego teoriia*. Moscow: Marakuev i Prianishnikov.

———. 1894b. Vitalism i nauka. *Russkaia mysl'* 11:155–69.

Todes, Daniel P. 1989. *Darwin without Malthus: The struggle for existence in Russian evolutionary thought*. New York: Oxford University Press.

———. 2002. *Pavlov's physiology factory: Experiment, interpretation, laboratory enterprise*. Baltimore: John Hopkins University Press.

———. 2007. "Pavlov's Communists." Unpublished manuscript.

Todes, Daniel P., and Nikolai Krementsov. 2010. Dialectical materialism and Soviet science in the 1920s and 1930s. In *A history of Russian thought*, ed. William Leatherbarrow and Derek Oxford, 340–67. Cambridge, England: Cambridge University Press.

Tolz, Vera. 1997. *Russian academicians and the revolution: Combining professionalism and politics*. London: Macmillan.

Torgovlia zhenskim molokom. 1925. *Krasnaia gazeta* (vechernii vypusk), May 14, 3.

Trimmer, Eric J. 1967. *Rejuvenation: The history of an idea*. London: Robert Hale.

Trudy VII Vsesoiuznogo s"ezda terapevtov, 25–31 maia 1925. 1926. Leningrad: n.p.

Tsalolikhin, S. Ia. 1993. Ernst Gekkel' v Rossii. *Voprosy istorii estestvoznaniia i tekhniki* 1:98–100.

Tsetlin, A. Ia. 1922. O perelivanii krovi. *Ginekologiia i akusherstvo* 1:112–18.

Tucker, Robert C. 1990. *Stalin in power: The revolution from above, 1928–1941*. New York: Norton.

Turney, Jon. 1995. Life in the laboratory: Public responses to experimental biology. *Public Understanding of Science* 4:153–76.

Twentyman, Jean M. G. 1971 *The organic vision of Helán Jaworski*. Richmond Hill, Surrey, England: New Atlantis Foundation.

U groba A. A. Bogdanova. 1928. *Pravda*, April 10, 5.

Umanskii, Vladislav. 1897. *Nevedomyi mir: Mars i ego zhiteli*. St. Petersburg: Ia. I. Liberman.

Ushakov, A. P. 1927. K voprosu o perelivanii krovi. *NKh* 4 (4): 327–30.

Vasilevskii, L. [V. L.]. 1923. O peresadke i omolazhivanii. *NiT* 1:9–11; 2:10–12.

———. 1924. Novoe ob operatsii omolozheniia. *Zvezda* 5:188–99.

———. 1925a. Noveishie opyty omolozheniia v Moskve. *VM*, September 11, 2.

———. 1925b. Novye opyty omolozheniia v SSSR. *VZ* 21–22:1275–78.

———. 1928a. Moskovskii Gos. Institut perelivaniia krovi. *Leningradskii meditsinskii zhurnal* 4:141–43.

———. 1928b. Perelivanie krovi. *VZ* 8:414–16.

———. 1928c. Prof. A. A. Bogdanov. *NiT* 17:5.

———. 1928d. Smert' na nauchnom postu. *VZ* 8:413.

———. 1928e. V Gosudarstvennom institute perelivaniia krovi. *NiT* 16:18.

V dome uchenykh. 1923. *Izvestiia*, November 13, 6.

Veil', S. S. 1929. V obshchestve vrachei materialistov. *VSM* 10:594–97.

V institute perelivaniia krovi. 1926. *VM*, December 21, 2.

Vlados, Kh. Kh., and N. M. Shustrov. 1927. *Klinicheskaia gematologiia*. Moscow and Leningrad: GIZ.

Vlados, Kh. Kh., and E. Tareev. 1924. Autogemoterapiia. *Terapevticheskii arkhiv* 2 (2): 199–208.

Voronoff, Serge. 1920. *Vivre! Etudes des moyens de relever l'énergie vitale et de prolonger la vie*. Paris: B. Grasset.

——— [Voronov]. 1924a. *Omolozhenie peresadkoi polovykh zhelez*. Leningrad: Prakticheskaia meditsina.

——— [Voronov]. 1924b. *Omolozhenie: Sorok tri privivki ot obez'iany k cheloveku*. Leningrad and Moscow: Kniga.

——— [Voronov]. 1924c. *Peresadka polovykh zhelez*. Khar'kov: Nauchnaia mysl'.

——— [Voronov]. 1924d. *Vivre*. Moscow: Sabashnikov.

——— [Voronov]. 1925. *Prakticheskoe primenenie peresadki organov v zhivotnovodstve*. Leningrad: Akademicheskoe izd-vo.

Voskresenskii, L. N. 1924. Opyty nabliudeniia nad "omolozheniem" liudei i krupnykh sel'skokhoziastvennykh zhivotnykh. In *Omolozhenie v Rossii 1924*, 98–126.

Vrach. 1929. Smert' na nauchnom postu. *VM*, March 21, 3.

V Sovnarkome RSFSR. 1926. *Izvestiia*, April 10, 2.

Vtoroi nauchnyi s"ezd vrachei Srednei Azii v Tashkente. 1925. *NKh* 3:420–21.

Vucinich, Alexander. 1976. *Social thought in tsarist Russia: The quest for a general science of society*. Chicago: University of Chicago Press.

———. 1984. *Empire of knowledge: The Academy of Sciences of the USSR, 1917–1970*. Berkeley: University of California Press.

———. 1988. *Darwin in Russian thought*. Berkeley: University of California Press.

Wasserstein, Alan G. 1996. Death and the internal milieu: Claude Bernard and the origins of experimental medicine. *Perspectives in Biology and Medicine* 39 (3): 313–26.

Weil, P.-Émile, and Paul Isch-Wall. 1925. *La transfusion du sang, étude biologique et clinique*. Paris: Masson.

Weindling, Paul. 1992. German-Soviet medical co-operation and the Institute for Racial Research, 1927–c. 1935. *German History* 10:177–206.

White, James D. 1981. Bogdanov in Tula. *Studies in Soviet Thought* 22 (1): 33–58.

Widdis, Emma. 2003. *Visions of a new land: Soviet film from the revolution to the Second World War*. New Haven, CT: Yale University Press.

Williams, Robert C. 1980. Collective immortality: The syndicalist origins of proletarian culture, 1905–1910. *Slavic Review* 39 (3): 389–402.

Winther, Rasmus G. 2001. August Weismann on germ-plasm variation. *JHB* 34 (3): 517–55.

Wortman, Richard. 1967. *The crisis of Russian populism*. London: Cambridge University Press. Paperback ed., New York: Cambridge University Press, 2008.

Young, J. H. 1964. James Blundell, experimental physiologist and obstetrician. *Medical History* 8 (2): 159–69.

Zalkind, A. E. 1925. O zabolevaniiakh partaktiva. *Krasnaia nov'* 4:187–203.

Zamiatin, E. 1922. *Gerbert Uells*. Petrograd: Epokha.

Zavadovskii, B. 1921. Problema starosti i omolozheniia v svete noveishikh rabot Shteinakha, Voronova i dr. *Krasnaia nov'* 3:146–76.

———. 1923. *Problema starosti i omolozheniia v svete ucheniia o vnutrennei sekretsii*. Moscow: Krasnaia nov'.

———. 1925. Darvinizm i lamarkizm i problema nasledovaniia priobretennykh priznakov. *Pod znamenem marksizma* 10–11:79–114.

Zelenogorskii, F. A. 1876–77. Iz istorii i teorii metodov issledovaniia i dokazatel'stva: O matematicheskom (goemetricheskom), metafizicheskom (intuitivnom), induktivnom i kriticheskom metodakh issledovaniia i dokazatel'stva. *Zapiski Imperatorskogo Khar'kovskogo Universiteta* 4 (1876): 1–48; 1 (1877): 49–160; 2 (1877): 161–245.

———. 1885. *Istoriia psikhologii ot Dekarta do nastoiashchego vremeni*. Kharkov: Tipografiia Universiteta.

———. 1998. *O metodakh issledovaniia i dokazatel'stva*. Moscow: ROSSPEN.

Zhitnikov, B. A. 1927. Uchenie o krovianykh gruppakh. *ZhDUV* 1:89–93.

Index

Page numbers in italics refer to illustrations.

174